Universitext: Tracts in Mathematics

Editorial Board
(North America):

J.H. Ewing
F.W. Gehring
P.R. Halmos

Universitext

Editors (North America): J.H. Ewing, F.W. Gehring, and P.R. Halmos

Aksoy/Khamsi: Nonstandard Methods in Fixed Point Theory
Aupetit: A Primer on Spectral Theory
Bachumikern: Linear Programming Duality
Benedetti/Petronio: Lectures on Hyperbolic Geometry
Berger: Geometry I, II (two volumes)
Bliedtner/Hansen: Potential Theory
Booss/Bleecker: Topology and Analysis
Carleson/Gamelin: Complex Dynamics
Cecil: Lie Sphere Geometry: With Applications to Submanifolds
Chandrasekharan: Classical Fourier Transforms
Charlap: Bieberbach Groups and Flat Manifolds
Chern: Complex Manifolds Without Potential Theory
Cohn: A Classical Invitation to Algebraic Numbers and Class Fields
Curtis: Abstract Linear Algebra
Curtis: Matrix Groups
van Dalen: Logic and Structure
Devlin: Fundamentals of Contemporary Set Theory
DiBenedetto: Degenerate Parabolic Equations
Dimca: Singularities and Topology of Hypersurfaces
Edwards: A Formal Background to Mathematics I a/b
Edwards: A Formal Background to Mathematics II a/b
Emery: Stochastic Calculus
Foulds: Graph Theory Applications
Frauenthal: Mathematical Modeling in Epidemiology
Fukhs/Rokhlin: Beginner's Course in Topology
Gallot/Hulin/Lafontaine: Riemannian Geometry
Gardiner: A First Course in Group Theory
Gårding/Tambour: Algebra for Computer Science
Godbillon: Dynamical Systems on Surfaces
Goldblatt: Orthogonality and Spacetime Geometry
Hiawka/Schoissengeier/Taschner: Geometric and Analytic Number Theory
Howe/Tan: Non-Abelian Harmonic Analysis: Applications of $SL(2,R)$
Humi/Miller: Second Course in Ordinary Differential Equations
Hurwitz/Kritikos: Lectures on Number Theory
Iversen: Cohomology of Sheaves
Jones/Morris/Pearson: Abstract Algebra and Famous Impossibilities
Kelly/Matthews: The Non-Euclidean Hyperbolic Plane
Kempf: Complex Abelian Varieties and Theta Functions
Kostrikin: Introduction to Algebra
Krasnoselskii/Pekrovskii: Systems with Hysteresis
Luecking/Rubel: Complex Analysis: A Functional Analysis Approach
MacLane/Moerdijk: Sheaves in Geometry and Logic
Marcus: Number Fields
McCarthy: Introduction to Arithmetical Functions
Meyer: Essential Mathematics for Applied Fields

(continued after index)

Joel H. Shapiro

Composition Operators

and Classical Function Theory

With 17 Illustrations

Springer-Verlag

New York Berlin Heidelberg London Paris
Tokyo Hong Kong Barcelona Budapest

Joel H. Shapiro
Department of Mathematics
Michigan State University
East Lansing, MI 48824
USA

Mathematics Subjects Classifications (1991): 30Cxx, 47B38

Library of Congress Cataloging-in-Publication Data
Shapiro, Joel H.
    Composition operators and classical function theory/Joel H. Shapiro
        p. cm. -- (Universitext. Tracts in mathematics)
    Includes bibliographical references and indexes.
    ISBN 0-387-94067-7 (alk. paper)
    1. Composition operators. 2. Geometric function theory.
    I. Title. II. Series.
    QA329.2.S48 1993
    515'.7246--dc20   93-26147

Printed on acid-free paper.

Production managed by Jim Harbison; manufacturing supervised by Jacqui Ashri.
Camera-ready copy prepared from the author's L^AT_EX files.
Printed and bound by R.R. Donnelley & Sons, Harrisonburg, Virginia.
Printed in the United States of America.

9 8 7 6 5 4 3 2 1

ISBN 0-387-94067-7 Springer-Verlag New York Berlin Heidelberg
ISBN 3-540-94067-7 Springer-Verlag Berlin Heidelberg New York

for Jane

Preface

The study of composition operators links some of the most basic questions you can ask about linear operators with beautiful classical results from analytic-function theory. The process invests old theorems with new meanings, and bestows upon functional analysis an intriguing class of concrete linear operators. Best of all, the subject can be appreciated by anyone with an interest in function theory or functional analysis, and a background roughly equivalent to the following twelve chapters of Rudin's textbook *Real and Complex Analysis* [Rdn '87]: Chapters 1–7 (measure and integration, L^p spaces, basic Hilbert and Banach space theory), and 10–14 (basic function theory through the Riemann Mapping Theorem).

In this book I introduce the reader to both the theory of composition operators, and the classical results that form its infrastructure. I develop the subject in a way that emphasizes its geometric content, staying as much as possible within the prerequisites set out in the twelve fundamental chapters of Rudin's book.

Although much of the material on operators is quite recent, this book is not intended to be an exhaustive survey. It is, quite simply, an invitation to join in the fun. The story goes something like this.

The setting is the simplest one consistent with serious "function-theoretic operator theory:" the unit disc U of the complex plane, and the Hilbert space H^2 of functions holomorphic on U with square summable power series coefficients. To each holomorphic function φ that takes U into itself we associate the *composition operator* C_φ defined by

$$C_\varphi f = f \circ \varphi \qquad (f \text{ holomorphic on the disc}),$$

and set for ourselves the goal of discovering the connection between the function theoretic properties of φ and the behavior of C_φ on H^2.

Indeed, it is already a significant accomplishment to show that each composition operator takes H^2 into itself; this is essentially Littlewood's famous Subordination Principle. Further investigation into properties like compactness, spectra, and cyclicity for composition operators leads naturally to classical results like the Julia-Carathéodory Theorem on the angular derivative, the Denjoy-Wolff iteration theorem, Königs's solution of Schröder's functional equation, the Koebe Distortion Theorem, and to hidden gems like the Linear Fractional Model Theorem, and Littlewood's "Counting Function" generalization of the Schwarz Lemma. I list below a more detailed outline.

Chapter 0. This is a prologue, to be consulted as needed, on the basic properties and classification of linear fractional transformations. Linear fractional maps play a vital role in our work, both as agents for changing coordinates and transforming settings (e.g., disc to half-plane), and as a source of examples that are easily managed, yet still rich enough to exhibit surprisingly diverse behavior. This diversity of behavior foreshadows the Linear Fractional Model Theorem, which asserts that *every* univalent self-map of the disc is conjugate to a linear fractional self-map of some, usually more complicated, plane domain.

Chapter 1. After developing some of the basic properties of H^2, we show that every composition operator acts boundedly on this Hilbert space. As pointed out above, this is essentially Littlewood's Subordination Principle. We present Littlewood's original proof—a beautiful argument that is perfectly transparent in its elegance, but utterly baffling in its lack of geometric insight. Much of the sequel can be regarded as an effort to understand the geometric underpinnings of this theorem.

Chapters 2 through 4: Having established the boundedness of composition operators, we seek to characterize those that are *compact*. Chapter 2 sets out the motivation for this problem, and in Chapter 3 we discover that the geometric soul of Littlewood's Theorem is bound up in the Schwarz Lemma. Armed with this insight, we are able to characterize the *univalently induced* compact composition operators, obtaining a compactness criterion that leads directly to the *Julia-Carathéodory Theorem* on the angular derivative. In Chapter 4 we prove the Julia-Carathéodory Theorem in a way that emphasizes its geometric content, especially its connection with the Schwarz Lemma. At the end of Chapter 4 we give further applications to the compactness problem.

Chapters 5 and 6: Closely related to the Julia-Carathéodory Theorem is the Denjoy-Wolff Iteration Theorem (Chapter 5), which plays a fundamental role in much of the further theory of composition operators. In Chapter 6 we use the Denjoy-Wolff Theorem as a tool in determining the *spectrum* of a compact composition operator. Our work on the spectrum leads to a connection between the Riesz Theory of compact operators and Königs's classical work on holomorphic solutions of Schröder's functional equation—the eigenvalue equation for composition operators. We develop the relevant part of the Riesz Theory, and show how, in helping to characterize the spectrum of a compact composition operator, it also provides a growth restriction on the solutions of Schröder's equation for the inducing map. When the inducing map is univalent this growth restriction has a compelling geometric interpretation, which re-emerges in Chapter 9 to provide, under appropriate hypotheses, a geometric characterization of compactness based on linear fractional models.

Chapter 7 and 8: The study of eigenfunctions begun in Chapter 6 leads in two directions. In these chapters we show how it suggests the idea of using simple linear fractional transformations as "models" for holomorphic self-maps of the disc. These *linear fractional models* provide an important tool for investigating other properties of composition operators. We illustrate their use in the study of *cyclicity*. Taking as inspiration earlier work of Birkhoff, Seidel, and Walsh, we show that certain composition operators exhibit *hypercyclicity;* they have a dense orbit (in the language of dynamics, they are *topologically transitive*). Our method involves characterizing the hypercyclic composition operators induced by linear fractional self-maps of U (Chapter 7), and then transferring these results, by means of an appropriate linear fractional model, to more general classes of inducing maps (Chapter 8).

Chapter 9: The second direction for the study of eigenfunctions involves turning around the growth condition discovered in Chapter 6 to provide a compactness criterion: If φ obeys reasonable hypotheses, and the solutions of Schröder's equation for φ do not grow too fast, then C_φ is compact. The requisite growth condition has a simple geometric interpretation in terms of the linear fractional model for φ. Again the key to this work is the Schwarz Lemma, this time in the form of hyperbolic distance estimates that follow from the Koebe Distortion Theorem.

Chapter 10: The solution to the compactness problem for general composition operators leads into value distribution theory, most notably the asymptotic behavior of the Nevanlinna Counting Function. Here our story comes full circle: the crucial inequality on the Counting Function is due to Littlewood, and when reinterpreted it becomes a striking generalization of the Schwarz Lemma.

As you might guess, the expository rules announced above occasionally get bent, but most of these transgressions can be remedied by referring to other parts of [Rdn '87]. The radial limit theorem for H^2 functions is introduced in Chapter 2, although my defense here is that its real application involves only the bounded case, which you find in the legally admissible part of [Rdn '87]. For Chapter 3 the reader is hoped to have acquired some of the lore concerning compactness and equicontinuity in metric spaces, and to have written down, at some point, the metric for uniform convergence on compact subsets of the disc. Chapters 6,8 and 10 utilize a standard convergence theorem for infinite products of holomorphic functions that occurs in the first few pages of [Rdn '87, Ch. 15].

The Carathéodory Theorem concerning boundary continuity of Riemann maps onto Jordan domains makes a couple of appearances; but note that it is "almost" proved in Chapter 14 of [Rdn '87]. The discussion of cyclicity for the "non-linear-fractional" case requires Walsh's theorem on polynomial approximation on Jordan domains, which is a special case of Mergelyan's Theorem (Chapter 20 of [Rdn '87]). Although the Linear Fractional Model Theorem is discussed in some generality, only those parts needed for the results presented here are proved.

Acknowledgments

This book originated from lectures I gave in March 1991 at Universidad Simon Bolivar in Caracas, and from a longer series of talks at the University of Hawaii spanning the Fall semester of that year. I am particularly grateful to Professor Julio Cano, who arranged my visit to Caracas, and to my colleagues at Hawaii who provided a lively and appreciative audience. In Honolulu I had the good fortune to interact with Wayne Smith and David Stegenga; and from our discussions emerged a joint project, part of which forms Chapter 9 of this book. During this period I also benefited from conversations with Tom Hoover, H.M. Hilden, Marvin Ortel, and H.S. Bear.

Over the past seven years I have had the privilege of collaborating on problems involving cyclicity and composition operators with Paul Bourdon of Washington and Lee University. The material of Chapters 7 and 8 is joint work with him, and my continuing interest in the subject owes a lot to his insight and enthusiasm.

My conversations with Bourdon, Smith, and Stegenga convinced me that the study of composition operators could serve as a showcase for the classical theorems that make the subject so appealing. Steven Krantz started me thinking about writing a book long before I felt ready to undertake it, and he ultimately provided the vital encouragement which got the project off the ground. Paul Bourdon and Wayne Smith read the original manuscript, caught many errors, and suggested numerous changes that greatly improved the exposition. I am deeply indebted to all these people for their support.

East Lansing, Michigan
April 1993

Contents

Preface   vii

Acknowledgments   x

Notation and Terminology   xv

0   Linear Fractional Prologue   **1**
0.1   First Properties   1
0.2   Fixed Points   2
0.3   Classification   4
0.4   Linear Fractional Self-Maps of U   5
0.5   Exercises   7

1   Littlewood's Theorem   **9**
1.1   The Hardy Space H^2   9
1.2 H^2 via Integral Means   11
1.3   Littlewood's Theorem   13
1.4   Exercises   17
1.5   Notes   19

2   Compactness: Introduction   **21**
2.1   Compact Operators   21
2.2   First Class of Examples   23
2.3   A Better Compactness Theorem   24
2.4   Compactness and Weak Convergence   29

2.5	Non-Compact Composition Operators	30
2.6	Exercises	33
2.7	Notes	35

3 Compactness and Univalence   **37**

3.1	The H^2 Norm via Area Integrals	37
3.2	The Theorem	38
3.3	Proof of Sufficiency	40
3.4	The Adjoint Operator	41
3.5	Proof of Necessity	43
3.6	Compactness and Contact	44
3.7	Exercises	51
3.8	Notes	52

4 The Angular Derivative   **55**

4.1	The Definition	56
4.2	The Julia-Carathéodory Theorem	57
4.3	The Invariant Schwarz Lemma	59
4.4	A Boundary Schwarz Lemma	61
4.5	Proof that (JC 1) $\Rightarrow$(JC 2)	65
4.6	Proof that (JC 2) $\Rightarrow$(JC 3)	69
4.7	Angular derivatives and contact	72
4.8	Exercises	72
4.9	Notes	75

5 Angular Derivatives and Iteration   **77**

5.1	Statement of Results	77
5.2	Elementary Cases	79
5.3	Wolff's Boundary Schwarz Lemma	81
5.4	Contraction Mappings	83
5.5	Grand Iteration Theorem, Completed	84
5.6	Exercises	85
5.7	Notes	86

6 Compactness and Eigenfunctions   **89**

6.1	Königs's Theorem	90
6.2	Eigenfunctions for Compact C_φ	93
6.3	Compactness vs. Growth of σ	96
6.4	Compactness vs. Size of $\sigma(U)$	97
6.5	Proof of Riesz's Theorem	99
6.6	Exercises	101
6.7	Notes	104

7   Linear Fractional Cyclicity   107
   7.1   Hypercyclic Fundamentals 108
   7.2   Linear Fractional Hypercyclicity 112
   7.3   Linear Fractional Cyclicity 118
   7.4   Exercises 124
   7.5   Notes 126

8   Cyclicity and Models   129
   8.1   Transference from Models 129
   8.2   From Maps to Models 132
   8.3   A General Hypercyclicity Theorem 133
   8.4   Exercises 141
   8.5   Notes 143

9   Compactness from Models   147
   9.1   Review of Königs's Model 147
   9.2   Motivation 148
   9.3   Main Result 149
   9.4   The Hyperbolic Distance on U 150
   9.5   The Hyperbolic Distance on G 153
   9.6   Twisted Sectors 158
   9.7   Main Theorem: Down Payment 160
   9.8   Three Lemmas 161
   9.9   Proof of the No-Sectors Theorem 166
   9.10 Exercises 170
   9.11 Notes 174

10 Compactness: General Case   177
   10.1 Motivation 177
   10.2 Inadequacy of Angular Derivatives 180
   10.3 Non-Univalent Changes of Variable 186
   10.4 Decay of the Counting Function 187
   10.5 Proof of Sufficiency 188
   10.6 Averaging the Counting Function 189
   10.7 Proof of Necessity 191
   10.8 Exercises 193
   10.9 Notes 195

Epilogue   199

References   203

Symbol Index   215

Author Index   217

Subject Index   220

Notation and Terminology

Here is a summary of the major notational and linguistic conventions used throughout the book.

Disc and Half-plane. The symbol U denotes the unit disc of the complex plane, often simply called "the disc," $\overline{U}$ denotes the closed unit disc, and ∂U the unit circle. The symbol φ always denotes a holomorphic self-mapping of U. Similarly, Π denotes the open right half-plane $\{w : \operatorname{Re} w > 0\}$, and $\overline{\Pi}$ is its closure in the plane.

The space of holomorphic functions. We denote by $H(U)$ the space of all holomorphic functions on U, and this space is always understood to be endowed with the topology κ of uniform convergence on compact subsets of U. The notation $f_n \overset{\kappa}{\to} f$ means that the sequence $\{f_n\}$ of functions converges to f uniformly on every compact subset of U.

For functions f holomorphic on U we adopt the notation $\hat{f}(n)$ for the n-th coefficient in the power series expansion of f about the origin:

$$f(z) = \sum_{n=0}^{\infty} \hat{f}(n) z^n \qquad (z \in U).$$

Hilbert space. "Hilbert space" always means "separable Hilbert space." The norm in any Hilbert space is denoted by " $\| \cdot \|$, " and the inner product by "$< \cdot, \cdot >$." The reader is assumed to be familiar with these Hilbert spaces: The usual Lebesgue space L^2, of (equivalence classes of) measurable functions on the the unit circle, square integrable with respect to arc-length measure; and ℓ^2, the space of square summable (one-sided) complex sequences.

Operators. The term *operator* always means *bounded linear transformation,* and *finite-rank operators* are bounded linear transformations with finite-dimensional range. When no confusion results, the symbol $\|\cdot\|$, originally used for the Hilbert space norm, will also be used for the norm of a bounded linear operator, or for any other norm that we want to discuss.

Special automorphisms. We will frequently employ the linear fractional transformation

$$\alpha_p(z) \overset{\text{def}}{=} \frac{p - z}{1 - \bar{p}z}$$

for $p \in U$. This is the *special automorphism* of U that interchanges the origin and the point p (see §4 of Chapter 0 below for more details).

Iterates. If φ is a holomorphic self-map of U and n is a positive integer, then the *n-th iterate* of φ is the n-fold composition of φ with itself. We always denote this map by φ_n:

$$\varphi_n = \varphi \circ \varphi \circ \cdots \circ \varphi \qquad (n \text{ times}),$$

taking care to warn the reader of any conflicts of notation that arise when iterates and other sequences populate the same argument.

Estimates. We frequently write estimates for non-negative functions that look like

$$A(x) \leq \text{const. } B(x),$$

for some range of x. In such inequalities the constant is always positive and finite, and is allowed to change from one occurrence to another, but it *never* depends on x. When we write

$$A(x) \approx B(x) \qquad (\text{for some range of } x)$$

we mean

$$\text{const. } A(x) \leq B(x) \leq \text{const. } A(x)$$

for the relevant values of x. Frequently occurring instances of this are the simple equivalences:

$$1 - x^2 \approx 1 - x \qquad (0 \leq x \leq 1),$$

and

$$1 - x \approx \log \frac{1}{x} \qquad (\text{as } x \to 1-).$$

0
Linear Fractional Prologue

Linear fractional transformations appear in this book at several different levels. Most frequently they play the traditional role of conformal changes of variable, allowing discs to be replaced by half-planes, and arbitrary points to be regarded as the origin. At a deeper level the linear fractional mappings that take the unit disc into itself provide a class of tractable examples diverse enough to reflect much of the character of more general situations. In fact, the Linear Fractional Model Theorem proclaims that *every univalent* self-map of the disc is conjugate (in a sense to be made precise in Chapters 6 and 8) to a linear fractional map.

This prologue sets out the fundamental classification of linear fractional transformations, as found early on in standard treatments like [Ahl '66], [Frd '29], and [Lnr '66]. Proofs of all but the most elementary results are sketched with enough detail so that the reader should have no difficulty in filling in the rest. I caution the reader against plowing through this chapter from beginning to end. It is intended as a reference section, to be consulted only as the need arises.

0.1   First Properties

A *linear fractional transformation* is a mapping of the form

$$T(z) = \frac{az + b}{cz + d},\qquad(1)$$

subject to the further condition $ad - bc \neq 0$, which is necessary and sufficient for T to be non-constant. We denote the set of all such maps by $LFT(\widehat{\mathbf{C}})$, where the notation is intended to call attention to the fact that, with the obvious conventions about the point at ∞, each linear fractional transformation can be regarded as a one-to-one holomorphic mapping of the Riemann Sphere $\widehat{\mathbf{C}}$ onto itself.

Group properties. $LFT(\widehat{\mathbf{C}})$ is a group under composition. Each of its members maps every circle on the sphere (i.e., every circle or line in the plane) to another circle, and given any pair of circles, some member of $LFT(\widehat{\mathbf{C}})$ maps one onto the other. The same is true for the set of triples of distinct points of the sphere. In the language of group theory:

> $LFT(\widehat{\mathbf{C}})$ acts transitively on the set of circles of $\widehat{\mathbf{C}}$, and triply transitively on the points.

Matrix representation. Each non-singular 2×2 complex matrix

$$A = \left(\begin{array}{cc} a & b \\ c & d \end{array} \right)$$

gives rise to a linear fractional transformation T_A by means of definition (1) above. Clearly, $T_A = T_{\lambda A}$ for any $\lambda \in \mathbf{C}$. For this reason it is convenient when working out general properties of $LFT(\widehat{\mathbf{C}})$ to normalize the matrices to have determinant $+1$.

Definition. If $ad - bc = 1$ in definition (1), we say T is in *standard form*.

Actually there are *two* standard forms, since the determinant is not changed when all coefficients are replaced by their negatives. Rather than introduce more definitions to remedy the situation, we will just keep this potential "plus-minus pitfall" in mind, and deal with it as the need arises.

The utility of matrices in dealing with linear fractional transformations comes from the fact that $T_A \circ T_B = T_{AB}$. Borrowing again from group theory, we say that $S, T \in LFT(\widehat{\mathbf{C}})$ are *conjugate* if there exists $V \in LFT(\widehat{\mathbf{C}})$ such that $S = V \circ T \circ V^{-1}$. Thus:

> *Conjugate linear fractional transformations correspond to similar matrices.*

0.2   Fixed Points

Clearly the linear fractional transformation $\frac{az+b}{cz+d}$ fixes the point ∞ if and only if $c = 0$, in which case ∞ is the *only* fixed point if and only if $a = d$ and $b \neq 0$. Otherwise the fixed point equation is a quadratic, with solutions

$$\alpha, \beta = \frac{(a-d) \pm \left[(a-d)^2 + 4bc\right]^{1/2}}{2c} .$$

The trace. If $T(z) = \frac{az+b}{cz+d}$ is in standard form ($ad - bc = 1$), then define the *trace* of T to be

$$\chi(T) = \pm(a + d),$$

where the ambiguous sign is intended to signal the "plus-minus" ambiguity in the standard form representation of T.

For example, T has ∞ as its only fixed point on the sphere, if and only if $T(z) = z + b$, in which case $|\chi(T)| = 2$. If T has only finite fixed points, then the equations written above for these fixed points can be at least partially expressed in terms of the trace:

$$\alpha, \beta = \frac{(a - d) \pm \left[\chi(T)^2 - 4\right]^{1/2}}{2c}. \tag{2}$$

This equation, plus our previous remark about maps with unique fixed point at ∞ shows that

$T \in LFT(\widehat{\mathbf{C}})$ has a unique fixed point in $\widehat{\mathbf{C}}$ if and only if $|\chi(T)| = 2$.

Derivatives at the fixed points. If $T \in LFT(\widehat{\mathbf{C}})$ is in standard form, then

$$T'(z) = \frac{ad - bc}{(cz + d)^2} = \frac{1}{(cz + d)^2},$$

and a little computation, using (2) above shows that the derivative of T at its fixed points can be represented in terms of the trace:

$$T'(\alpha),\ T'(\beta) = \frac{1}{4}\left(\chi(T) \pm \left[\chi(T)^2 - 4\right]^{1/2}\right)^2, \tag{3}$$

where the ambiguity in the sign of the trace is absorbed by the fact that the right-hand side is a perfect square. From this it follows readily that

$$T'(\alpha) = \frac{1}{T'(\beta)} \quad \text{and} \quad T'(\alpha) + T'(\beta) = \chi(T)^2 - 2. \tag{4}$$

In case T has a fixed point at ∞ and another finite one, it must have the form $T(z) = az + b$, in which case we define $T'(\infty) = (1/T)'(0)^{-1} = 1/a$. Hence, once again the relations (4) hold. In summary:

Theorem (Fixed points and derivatives). *Suppose $T \in LFT(\widehat{\mathbf{C}})$. Then these are equivalent:*

- $|\chi(T)| = 2$.

- $T' = 1$ at a fixed point of T.

- T has just one fixed point on $\widehat{\mathbf{C}}$.

If T has two distinct fixed points, then its derivatives at these points are reciprocals, and their sum is $\chi(T)^2 - 2$.

0.3   Classification

A map $T \in LFT(\widehat{\mathbf{C}})$ is called *parabolic* if it has a single fixed point in $\widehat{\mathbf{C}}$. Suppose T is parabolic and has its fixed point at $\alpha \in \mathbf{C}$. If $S \in LFT(\widehat{\mathbf{C}})$ takes α to ∞, then $V = S \circ T \circ S^{-1}$ belongs to $LFT(\widehat{\mathbf{C}})$ and fixes only the point ∞. Therefore $V(z) = z + \tau$ for some non-zero complex number τ. Thus, every parabolic linear fractional map is conjugate to a translation.

If T is not parabolic, there are two fixed points $\alpha, \beta \in \widehat{\mathbf{C}}$. Let S be a linear fractional map that takes α to 0 and β to ∞. Then the map $V = S \circ T \circ S^{-1}$ belongs to $LFT(\widehat{\mathbf{C}})$ and fixes both 0 and ∞, so it must have the form $V(z) = \lambda z$ for some complex number λ, which is called the *multiplier* for T. Thus:

$$T(z) = S^{-1}(\lambda S(z)) \qquad (z \in \mathbf{C}). \tag{5}$$

By the chain rule,

$$T'(\alpha) = \lambda \quad \text{and} \quad T'(\beta) = \frac{1}{\lambda}, \tag{6}$$

(this also follows from (5) and the theorem of the last section). Equation (6) insures that, if $|\lambda| \neq 1$, then one of the fixed points of T is *attractive*. For example, if $|\lambda| < 1$ this attractive fixed point is α, and

$$T_n(z) \to \alpha \qquad (z \in \widehat{\mathbf{C}} \backslash \{\beta\}),$$

where T_n denotes the n-th iterate of T (see *Notation and Terminology*), and the convergence is even uniform on compact subsets o. $\widehat{\mathbf{C}} \backslash \{\beta\}$.

Note that there is also ambiguity in the definition of multiplier. If the roles of α and β are interchanged, so that now S sends β to zero, then $1/\lambda$ is the multiplier. Thus it is really the *pair* $\{\lambda, 1/\lambda\}$ that should be regarded as the multiplier of the transformation. The theorem on fixed points and derivatives, along with (6) above, shows that

$$\lambda + \frac{1}{\lambda} = \chi(T)^2 - 2, \tag{7}$$

a relationship that takes into account all ambiguities in the definition of both trace and multiplier.

Multipliers furnish the following classification of non-parabolic maps.

Definition. Suppose $T \in LFT(\widehat{\mathbf{C}})$ is neither parabolic nor the identity. Let $\lambda \neq 1$ be the multiplier of T. Then T is called:

- *Elliptic* if $|\lambda| = 1$,

- *Hyperbolic* if $\lambda > 0$, and

- *Loxodromic* if T is neither elliptic nor parabolic.

Thus, the classification of linear fractional transformations that are not the identity can be summarized as follows:

> Parabolic maps are conjugate to translations, elliptic maps to rotations, hyperbolic ones to positive dilations, and all others (the loxodromic maps) are conjugate to complex dilations.

By (7) this classification can also be expressed in terms of the trace.

Theorem (Classification by the trace). *Suppose T is a linear fractional map that is not the identity. Then T is loxodromic if and only if its trace $\chi(T)$ is not real. If $\chi(T)$ is real, then T is:*

- *hyperbolic $\iff |\chi(T)| > 2$,*

- *parabolic $\iff |\chi(T)| = 2$,*

- *elliptic $\iff |\chi(T)| < 2$.*

0.4 Linear Fractional Self-Maps of U

Our interest here is in $LFT(U)$, the subgroup of $LFT(\widehat{\mathbf{C}})$ consisting of self-maps of the unit disc U. Those that take U onto itself are called *(conformal) automorphisms*. Consideration of normal forms quickly shows that:

(a) Parabolic members of $LFT(U)$ have their fixed point on ∂U.

(b) Hyperbolic members of $LFT(U)$ must have attractive fixed point in $\overline{U}$, with the other fixed point outside U, and lying on ∂U if and only if the map is an automorphism of U.

(c) Loxodromic and elliptic members of $LFT(U)$ have a fixed point in U and a fixed point outside $\overline{U}$. The elliptic ones are precisely the automorphisms in $LFT(U)$ with this fixed point configuration.

The proof is straightforward, and we omit it, except to say that in considering the normal forms, the unit disc gets replaced by a disc or half-plane that is taken into itself under the map in question, and from this invariance one quickly deduces the properties listed above.

We next consider two sub-classes of $LFT(U)$ that will be of crucial importance.

Automorphisms of U. We denote the class of automorphisms of U by Aut(U). Particularly useful automorphisms are the *special* ones α_p, defined for each $p \in U$ by

$$\alpha_p(z) \stackrel{\text{def}}{=} \frac{p - z}{1 - \bar{p}z}.$$

The map α_p interchanges the point p and the origin, and has the advantage of being its own compositional inverse. Moreover, the Schwarz Lemma shows that every automorphism of U is a composition of a standard one and a rotation (see Exercise 2). We will frequently use the special automorphisms to effect conformal changes of variable, thus allowing any particular point of the disc to temporarily serve as the origin.

Maps in $LFT(U)$ with no interior fixed point. Certain aspects of our work focus on holomorphic self-maps of U with no fixed point in U, and in particular on linear fractional maps with this property. As we saw above, a linear fractional self-map of U with no interior fixed point can only be hyperbolic or parabolic, and its attractive fixed point must lie on ∂U. Suppose ψ is such a map, and suppose for simplicity that its attractive fixed point is $+1$. Upon conjugating ψ by the transformation $w = (1+z)/(1-z)$ we arrive at a linear fractional map Ψ that takes the right half-plane Π into itself, and fixes ∞; hence

$$\Psi(w) = \lambda w + a, \tag{8}$$

where the fact that Ψ preserves the right half plane forces $\lambda > 0$ and $\operatorname{Re} a \geq 0$. Since we are making ∞ the attractive fixed point of Ψ, we actually have $\lambda \geq 1$, with equality if and only if Ψ, and therefore ψ, is parabolic. Upon changing variables in this "half-plane model" for ψ, we arrive at this formula:

$$\psi(z) = 1 + \frac{2(z-1)}{2\lambda + (\lambda - a - 1)(z-1)}, \tag{9}$$

from which a simple computation yields $\psi'(1) = 1/\lambda$ (which we already know from previous work), and

$$\psi''(1) = \frac{a + 1 - \lambda}{\lambda^2}.$$

If ψ is parabolic, then $\lambda = 1$, so ψ's *alter ego* in the right half-plane is the translation map $\Psi(w) = w + a$, and this is an automorphism of Π if and only if a is pure imaginary. Thus the parabolic automorphisms and non-automorphisms of the disc can be distinguished from each other by the following second-derivative criterion.

Proposition. *Suppose $\psi \in LFT(U)$ is parabolic and fixes the point $+1$. Then:*

(a) *$\psi''(1)$ has non-negative real part, and is $\neq 0$.*

(b) *$\operatorname{Re} \psi''(1) = 0$ if and only if ψ is an automorphism of U.*

Finally, we note for emphasis that if $\psi \in LFT(U)$ is parabolic with fixed point at $+1$, then the explicit formula for ψ is (9), with $\lambda = 1$:

$$\psi(z) = 1 + \frac{2(z-1)}{2 - a(z-1)} \quad \text{where} \quad \operatorname{Re} a \geq 0. \tag{10}$$

We will seldom embrace explicit formulas for linear fractional maps, but this one is an exception; it will play an important role in the work of Chapter 7.

0.5   Exercises

1. Show that the following maps take U into itself, determine the position of their fixed points, and classify each map.

   (a) $\dfrac{1}{2-z}$,   (b) $\dfrac{z}{2-z}$,   (c) $\dfrac{-z}{2+z}$,   (d) $\dfrac{1+z}{2}$.

2. Show that every automorphism of U has the form $\omega \alpha_p$ for some $p \in U$ and complex number ω of modulus 1.

   Suggestion: For T an automorphism, let $p = T^{-1}(0)$. Consider $T \circ \alpha_p$.

   Can every automorphism be represented as $\alpha_p(\omega z)$?

3. Show that for each $p \in U$ the special automorphisms are all elliptic, but upon multiplying by -1 they all become hyperbolic.

4. More generally, suppose $\omega = e^{i\theta_0}$ where $-\pi < \theta_0 \leq \pi$, and that $\varphi \subset U$. Show that $\omega \alpha_p$ is:

   - elliptic $\iff |p| < \cos(\theta_0/2)$,
   - parabolic $\iff |p| = \cos(\theta_0/2)$,
   - hyperbolic $\iff |p| > \cos(\theta_0/2)$.

5. Suppose $T \in LFT(\widehat{\mathbf{C}})$ is not parabolic, and has multiplier λ. Use normal forms and similarity of matrices to derive the following relations between multiplier and trace (the last one is formula (7) of the text):

   (a) $\chi(T) = \lambda^{1/2} + \lambda^{-1/2}$,
   (b) $\chi(T)^2 = \lambda + \lambda^{-1} - 2$.

6. Suppose $\psi \in LFT(U)$ is parabolic with fixed point at $+1$. As pointed out in §0.4, the statements below follow easily from the fact that the corresponding "right half-plane model" of ψ is $\Psi(w) = w + a$:

(a) $\operatorname{Re} a \geq 0$.

(b) $\operatorname{Re} a = 0 \iff \psi \in \operatorname{Aut}(U)$.

Show that these statements can be derived instead from the fact that the boundary of $\psi(U)$ has curvature ≥ 1, with equality if and only if ψ is an automorphism.

1
Littlewood's Theorem

We introduce the Hilbert space H^2 of analytic functions, discuss its norm, and give Littlewood's original proof that every composition operator takes H^2 boundedly into itself.

1.1 The Hardy Space H^2

We say that a function

$$f(z) = \sum_{n=0}^{\infty} \hat{f}(n)z^n \in H(U) \tag{1}$$

belongs to the *Hardy space* H^2 if its sequence of power series coefficients is square-summable:

$$H^2 = \{f \in H(U) : \sum_{n=0}^{\infty} |\hat{f}(n)|^2 < \infty\}.$$

Note that *every* square-summable sequence of complex numbers is the coefficient sequence of an H^2 function; if $\{a_n\}_0^\infty$ is square-summable, then it is bounded, so the corresponding power series $\sum a_n z^n$ converges on U to an analytic function that belongs to H^2. By the uniqueness theorem for power series, the map that associates the function f with the sequence $\{\hat{f}(n)\}$ is therefore a vector space isomorphism of H^2 onto ℓ^2, the Hilbert space of square summable complex sequences, and we turn H^2 into a Hilbert space by declaring this map to be an isometry:

$$\|f\| = \left(\sum_{n=0}^{\infty} |\hat{f}(n)|^2 \right)^{1/2} \qquad (f \in H^2).$$

Some properties of H^2 follow readily from this definition. The next result shows, for example, that H^2 functions cannot grow too rapidly.

Growth Estimate. *For each $f \in H^2$,*

$$|f(z)| \le \frac{\|f\|}{\sqrt{1 - |z|^2}} \tag{2}$$

for each $z \in U$.

Proof. Upon applying the Cauchy-Schwarz inequality to the power series representation (1) of f we obtain for each $z \in U$,

$$|f(z)| \le \sum_{n=0}^{\infty} |\hat{f}(n)||z|^n \le \left(\sum_{n=0}^{\infty} |\hat{f}(n)|^2 \right)^{\frac{1}{2}} \left(\sum_{n=0}^{\infty} |z|^{2n} \right)^{\frac{1}{2}},$$

and the rest follows by using the definition of norm in H^2 and summing the geometric series on the right.   □

This inequality shows that the topology of H^2 is, in the following sense, "natural" for analytic functions.

Corollary. *Every norm convergent sequence in H^2 converges (to the same limit) uniformly on compact subsets of U.*

Proof. Suppose $\{f_n\}$ is a sequence in H^2 norm-convergent to a function $f \in H^2$, that is, $\|f_n - f\| \to 0$. For $0 < R < 1$ the growth estimate above yields for each fixed n,

$$\sup_{|z| \le R} |f_n(z) - f(z)| \le \frac{\|f_n - f\|}{\sqrt{1 - R^2}},$$

so $f_n \to f$ uniformly on the closed disc $\{|z| \le R\}$. Since R is arbitrary, $f_n \to f$ uniformly on every compact subset of U.   □

It is also easy to see from the definition that H^2 contains some unbounded functions. For example,

$$\log \frac{1}{1 - z} = \sum_{n=1}^{\infty} \frac{z^n}{n} \in H^2.$$

With a little more work, you can use the binomial theorem to show that the function $(1-z)^{-\alpha} \in H^2$ for every $0 < \alpha < 1/2$, which establishes that the exponent in the Growth Estimate is best possible.

However the definition of H^2 in terms of coefficients more often obscures than reveals. Here, for example, are two important facts it does *not* elucidate (the symbol H^∞ denotes the space of bounded analytic functions on U).

- $H^\infty \subset H^2$. More generally, if $b \in H^\infty$ and $f \in H^2$ then the pointwise product $bf \in H^2$.

- ("Littlewood's Theorem.") If φ is a holomorphic self-map of U, then $f \circ \varphi \in H^2$ for every $f \in H^2$.

Both statements say something about linear operators. The first one asserts that the operator of "multiplication by b,"

$$M_b f = bf \qquad (f \in H^2)$$

takes H^2 into itself, while the result we are calling Littlewood's Theorem says the same thing about the *composition operator* C_φ:

$$C_\varphi f = f \circ \varphi \qquad (f \in H^2).$$

To see what is at stake here, suppose we try to prove the first statement, armed only with the definition of H^2. It is easy to express the power series coefficients of bf as a convolution of those of b and f, but then the argument founders on our inability to characterize the power series coefficients of bounded analytic functions.

A similar attempt to prove Littlewood's Theorem fares even worse, for now in the power series representation (1) we have to replace z^n by $\varphi(z)^n$, expand each power of φ using the binomial theorem, group like powers of z together, and then, knowing no more than before about the coefficients of φ, show that the resulting sequence of coefficients is square summable!

Thus, significant progress on most questions involving H^2 functions requires that the norm of the space be expressed in a way that emphasizes the *values* of analytic functions rather than their power series coefficients.

1.2 H^2 via Integral Means

Suppose

$$f(z) = \sum_{n=0}^{\infty} \hat{f}(n) z^n$$

is a function holomorphic on U. Upon writing $z = re^{i\theta}$, and using the orthogonality of the functions $\{e^{in\theta}\}_0^\infty$ in L^2, we see that for $0 \leq r < 1$,

$$M_2^2(f,r) \overset{\text{def}}{=} \frac{1}{2\pi} \int_{-\pi}^{\pi} |f(re^{i\theta})|^2 d\theta = \sum_{n=0}^{\infty} |\hat{f}(n)|^2 r^{2n} \tag{3}$$

This formula shows that $M_2(f,r)$ is an increasing function of r, and if $f \in H^2$, then it is bounded by the H^2 norm of f. Conversely, if

$$\lim_{r \to 1-} M_2(f,r) = M < \infty,$$

then for each non-negative integer N we have

$$\sum_{n=0}^{N} |\hat{f}(n)|^2 r^{2n} \leq \sum_{n=0}^{\infty} |\hat{f}(n)|^2 r^{2n} \leq M^2.$$

Upon sending r to 1 we see that each partial sum of the series for $\|f\|^2$ is bounded by M^2, hence the same is true of the whole series. Thus $f \in H^2$ and $\|f\| \leq M$. This completes the derivation of the first of our alternate expressions for the H^2 norm. To get a cleaner summary let us agree (from now on) to write $\|f\| = \infty$ whenever $f \notin H^2$.

Proposition. *Suppose f is holomorphic on U. Then as $r \nearrow 1$ the mean $M_2(f,r)$ increases to $\|f\|$:*

$$\|f\|^2 = \lim_{r \to 1-} \frac{1}{2\pi} \int_{-\pi}^{\pi} |f(re^{i\theta})|^2 dt, \tag{4}$$

Thus $f \in H^2$ if and only if $M_2(f,r)$ is bounded for $0 \leq r < 1$.

To test the utility of these results, let us return to the two important facts left hanging at the end of the last section. For $b \in H^\infty$ write

$$\|b\|_\infty = \sup_{z \in U} |b(z)|.$$

Since integrals of larger functions are larger, we see, with no effort at all, that $M_2(b,r) \leq \|b\|_\infty$ for all $0 < r < 1$, and, therefore, $b \in H^2$ with $\|b\|_\infty \leq \|b\|$. It is no more trouble to prove the more general statement that for each $f \in H^2$, the pointwise product bf also belongs to H^2, with a corresponding norm inequality:

$$\|bf\| \leq \|b\|_\infty \|f\| \qquad (b \in H^\infty, f \in H^2). \tag{5}$$

1.3 Littlewood's Theorem

O r integral representation of the H^2 norm paves the way for a proof of
Littlewood's Theorem. At first glance this optimism might seem premature,
since the application of formula (4) to $C_\varphi f = f \circ \varphi$ raises the specter of a
potentially unappetizing change of variable, possibly involving unbounded
derivatives, or multiple coverings, or both.

Fortunately, none of this discouraged Littlewood from using the tools
at hand to construct the beautiful proof we are going to present. Everything revolves around the special case $\varphi(0) = 0$, after which the full result
follows by means of a manageable conformal change of variable. The case
$\varphi(0) = 0$, which is the only one Littlewood actually considered, furnishes
two surprises. First, the proof requires only the fact that, since φ maps U
into itself, the *multiplication operator* M_φ acts *contractively* on H^2,

$$\|M_\varphi f\| \leq \|f\| \quad (f \in H^2) \tag{6}$$

(from (5) above). Second, this contractive property of M_φ gets passed on
to C_φ!

Littlewood's Subordination Principle (1925). *Suppose φ is a holomorphic self-map of U, with $\varphi(0) = 0$. Then for each $f \in H^2$,*

$$C_\varphi f \in H^2 \quad \text{and} \quad \|C_\varphi f\| \leq \|f\|.$$

Remark Recall that a linear operator T on a Hilbert space $\mathcal{H}$ is said to be
bounded if it takes the unit ball $\mathcal{B}$ into a bounded set, and that the *norm*
of such an operator is defined to be

$$\|T\| = \sup\{\|Tf\| : f \in \mathcal{B}\}.$$

In this language, the work of the last section asserts that if $b \in H^\infty$ then the
operator of multiplication by b is bounded on H^2, and has norm $\leq \|b\|_\infty$.

The bounded linear operators on a Hilbert space $\mathcal{H}$ are precisely the
continuous ones ([Rdn '87], Theorem 5.4, page 96). If $\|Tf\| \leq \|f\|$ for
each $f \in \mathcal{H}$ (i.e., if $\|T\| \leq 1$), then T is called a *contraction* on $\mathcal{H}$. Thus
the multiplication operator M_φ is a contraction on H^2, and Littlewood's
Subordination Principle asserts that the same is true of the composition
operator C_φ whenever φ fixes the origin. Stated this way, Littlewood's
Principle becomes the first (and most necessary) result in our program of
matching the properties of φ with the behavior of C_φ.

Proof. The proof is organized by the *backward shift operator* B, defined
on H^2 by

$$Bf(z) = \sum_{n=0}^{\infty} \hat{f}(n+1)z^n \quad (f \in H^2).$$

The name comes from the fact that B shifts the power series coefficients of f one unit to the left, and drops off the constant term. Clearly, B is a contraction on H^2, and one might expect this fact to play an important role in the proof. Surprisingly, it does not—not even the boundedness of B is required. Only the following two identities are needed, and they hold for any $f \in H(U)$:

$$f(z) = f(0) + zBf(z) \qquad (z \in U), \tag{7}$$

$$B^n f(0) = \hat{f}(n) \qquad (n = 0, 1, 2, \ldots). \tag{8}$$

To begin the proof, suppose first that f is a (holomorphic) polynomial. Then $f \circ \varphi$ is bounded on U, so by the work of the last section there is no doubt that it lies in H^2; the real issue is its *norm*.

We begin the norm estimate by substituting $\varphi(z)$ for z in (7), to obtain

$$f(\varphi(z)) = f(0) + \varphi(z)(Bf)(\varphi(z)) \qquad (z \in U),$$

which we rewrite in the language of composition and multiplication operators as

$$C_\varphi f = f(0) + M_\varphi C_\varphi B f. \tag{9}$$

At this point, the assumption $\varphi(0) = 0$ makes its first (and only) appearance. It asserts that all the terms of the power series for φ have a common factor of z, hence the same is true for the second term on the right side of equation (9), rendering it orthogonal in H^2 to the constant function $f(0)$. Thus,

$$\|C_\varphi f\|^2 = |f(0)|^2 + \|M_\varphi C_\varphi B f\|^2 \le |f(0)|^2 + \|C_\varphi B f)\|^2, \tag{10}$$

where the last inequality follows from (6) above, the contractive property of M_φ. Now successively substitute $Bf, B^2 f, \cdots$ for f in (10) to obtain:

$$\|C_\varphi B f\|^2 \le |Bf(0)|^2 + \|C_\varphi B^2 f\|^2$$

$$\|C_\varphi B^2 f\|^2 \le |B^2 f(0)|^2 + \|C_\varphi B^3 f\|^2$$

$$\vdots \qquad\qquad \vdots$$

$$\|C_\varphi B^n f\|^2 \le |B^n f(0)|^2 + \|C_\varphi B^{n+1} f\|^2.$$

Putting all these inequalities together, we get

$$\|C_\varphi f\|^2 \le \sum_{k=0}^{n} |(B^k f)(0)|^2 + \|C_\varphi B^{n+1} f\|^2$$

for each non-negative integer n.

Now recall that f is a polynomial. If we choose n be the degree of f, then $B^{n+1}f = 0$, and this reduces the last inequality to

$$\|C_\varphi f\|^2 \leq \sum_{k=0}^{n} |(B^k f)(0)|^2$$

$$= \sum_{k=0}^{n} |\hat{f}(k)|^2$$

$$= \|f\|^2,$$

where the middle line comes from property (8) of the backward shift. This shows that C_φ is an H^2-norm contraction, at least on the vector space of holomorphic polynomials.

For the endgame, suppose $f \in H^2$ is not a polynomial. Let f_n be the n-th partial sum of its Taylor series (1). Then $f_n \to f$ in the norm of H^2, so as we showed in §1.1, $f_n \overset{\kappa}{\to} f$, hence $f_n \circ \varphi \overset{\kappa}{\to} f \circ \varphi$. It is clear that $\|f_n\| \leq \|f\|$, and we have just shown that $\|f_n \circ \varphi\| \leq \|f_n\|$. Thus for each fixed $0 < r < 1$ we have (recalling the abbreviation $M_2(f, r)$ for the L^2 norm of f over the circle of radius r)

$$M_2(f \circ \varphi, r) = \lim_{n \to \infty} M_2(f_n \circ \varphi, r) \quad \text{(because } f_n \circ \varphi \overset{\kappa}{\to} f \circ \varphi\text{)}$$

$$\leq \limsup_{n \to \infty} \|f_n \circ \varphi\|$$

$$\leq \limsup_{n \to \infty} \|f_n\|$$

$$\leq \|f\|.$$

To complete the proof, let r tend to 1, and appeal to equation (4).   □

Automorphism-induced composition operators. To prove that C_φ is bounded even when φ does not fix the origin, we utilize conformal automorphisms to move points of U from where they are to where we want them. For each point $p \in U$, recall the *special automorphism*

$$\alpha_p(z) = \frac{p - z}{1 - \bar{p}z}, \tag{11}$$

that takes U onto itself, interchanges p with the origin, and is its own inverse. Write $p = \varphi(0)$. Then the holomorphic function $\psi = \alpha_p \circ \varphi$ takes U into itself and fixes the origin. By the self-inverse property of α_p we have $\varphi = \alpha_p \circ \psi$, and this translates into the operator equation $C_\varphi = C_\psi C_{\alpha_p}$. We have just seen that C_ψ is bounded (a contraction, in fact), and we know that the product of bounded operators is always bounded. Thus, the boundedness of C_φ on H^2 will follow from the first sentence of the next result.

Lemma. *For each $p \in U$ the operator C_{α_p} is bounded on H^2. Moreover,*

$$\|C_{\alpha_p}\| \leq \left(\frac{1+|p|}{1-|p|}\right)^{\frac{1}{2}}.$$

Proof. Suppose first that f is a holomorphic in a neighborhood of the closed unit disc, say in RU for some $R > 1$. Then the limit in formula (4) can be passed inside the integral sign, with the result that

$$\|f\|^2 = \frac{1}{2\pi}\int_{-\pi}^{\pi}|f(e^{i\theta})|^2 d\theta.$$

This opens the door to a simple change of variable:

$$
\begin{aligned}
\|f \circ \alpha_p\|^2 &= \frac{1}{2\pi}\int_{-\pi}^{\pi}|f(\alpha_p(e^{i\theta}))|^2 d\theta \\[2mm]
&= \frac{1}{2\pi}\int_{-\pi}^{\pi}|f(e^{it})|^2 |\alpha_p'(e^{it})| dt \\[2mm]
&= \frac{1}{2\pi}\int_{-\pi}^{\pi}|f(e^{it})|^2 \frac{1-|p|^2}{|1-\bar{p}e^{it}|^2} dt \\[2mm]
&\leq \frac{1-|p|^2}{(1-|p|)^2} \cdot \left(\frac{1}{2\pi}\int_{-\pi}^{\pi}|f(e^{it})|^2 dt\right) \\[2mm]
&= \frac{1+|p|}{1-|p|} \cdot \|f\|^2.
\end{aligned}
$$

Thus the desired inequality holds for all functions holomorphic in RU; in particular it holds for polynomials. It remains only to transfer the result to the rest of H^2, and for this we simply repeat the argument used to finish the proof of Littlewood's Subordination Theorem.   □

At this point we have assembled everything we need to show that composition operators act boundedly on H^2.

Littlewood's Theorem. *Suppose φ is a holomorphic self-map of U. Then C_φ is a bounded operator on H^2, and*

$$\|C_\varphi\| \leq \sqrt{\frac{1+|\varphi(0)|}{1-|\varphi(0)|}}.$$

Proof. As outlined earlier, we have $C_\varphi = C_\psi C_{\alpha_p}$, where $p = \varphi(0)$, and ψ fixes the origin. The last lemma and Littlewood's Subordination Principle show that both operators on the right are bounded on H^2, hence C_φ is the product of bounded operators on H^2, and is therefore itself bounded. Moreover

$$\|C_\varphi\| \le \|C_\psi\| \|C_{\alpha_p}\| \le \sqrt{\frac{1 + |\varphi(0)|}{1 - |\varphi(0)|}}$$

where the last inequality follows from the lemma and the fact that C_ψ is a contraction.   □

1.4   Exercises

1. Show that for every $0 < \alpha < 1/2$, the function $\left(\frac{1+z}{1-z}\right)^\alpha$ belongs to H^2.

2. Suppose f and g are holomorphic on U, with g univalent and $f(U) \subset g(U)$. Show that if g belongs to H^2, then so does f. *Suggestion*: Use Littlewood's Theorem.

3. Use the results of the previous two exercises to show that if f is holomorphic on U, and $f(U)$ is contained in an angular sector with vertex angle less than $\pi/2$ radians, then $f \in H^2$.

4. Here is another setting for the theory of composition operators. The *Bergman space* A^2 is the collection of functions $f \in H(U)$ for which

$$\|f\|^2 \stackrel{\text{def}}{=} \sum_{n=0}^\infty \frac{|\hat{f}(n)|^2}{n+1} < \infty.$$

Thus $H^2 \subset A^2$. Show that for every $f \in H(U)$:

$$\|f\|^2 = \int_U |f(z)|^2 \, dA(z),$$

where dA is Lebesgue area measure on U, normalized to have total mass 1 (i.e., $dA = \frac{1}{\pi} dx dy$), and "infinite norm" means $f \notin A^2$.

5. Show that every composition operator is bounded on A^2 (you can use the proof of Littlewood's Theorem, or the boundedness result for H^2 along with the integral form of the Bergman norm).

The next four problems examine the notion of composition operator on the
Dirichlet space $\mathcal{D}$, which is the collection of functions $f \in H(U)$ for which

$$D(f) \overset{\text{def}}{=} \sum_{n=1}^{\infty} n|\hat{f}(n)|^2 < \infty.$$

Thus $\mathcal{D} \subset H^2$.

6. Show that $f \in H(U)$ belongs to the Dirichlet space if and only if

$$\int_U |f'(z)|^2 \, dA(z) < \infty.$$

   Express $D(f)$ in terms of this integral, and in case f is univalent,
   relate the integral to the area of $f(U)$.

7. Show that there are composition operators that do not take $\mathcal{D}$ into
   itself.

   Suggestion: In order to take $\mathcal{D}$ into itself, the operator must be induced by a function in $\mathcal{D}$. This raises the question of whether every
   holomorphic self map φ of U that lies in $\mathcal{D}$ induces a bounded composition operator on $\mathcal{D}$. The answer is still "no," but it is much more
   difficult to produce an example.

8. Turn $\mathcal{D}$ into a Hilbert space by defining:

$$\|f\|_{\mathcal{D}}^2 = |f(0)|^2 + D(f) \qquad (f \in \mathcal{D}).$$

   Show that the operator M_z of "multiplication by z" is not a contraction on $\mathcal{D}$.

9. Show that, nevertheless, every *univalent* holomorphic self-map of U
   induces a bounded composition operator on $\mathcal{D}$.

10. Let $\ell^1(U)$ denote the space of absolutely summable complex sequen-
    ces, but now regarded as a space of analytic functions. That is,

$$\ell^1(U) = \{f \in H(U) : \|f\|_1 \overset{\text{def}}{=} \sum_{n=0}^{\infty} |\hat{f}(n)| < \infty\}$$

    Show that the following version of Littlewood's Subordination Theo-
    rem holds for $\ell^1(U)$:

    > Suppose $\varphi \in \ell^1(U)$ and $\|\varphi\|_1 \leq 1$. Then φ is a holomorphic
    > self-map of U, and C_φ is a contraction on $\ell^1(U)$.

    (See the *Notes* below for more on this problem.)

11. Suppose φ is a holomorphic self-map of U that belongs to $\ell^1(U)$, and $M > 0$. Show that the following are equivalent:

   (a) C_φ maps $\ell^1(U)$ into itself, with $\|C_\varphi f\|_1 \leq M\|f\|_1$ for each $f \in \ell^1(U)$.

   (b) $\|\varphi^n\|_1 \leq M$ for each positive integer n.

   Use this result to prove that $\varphi(z) = (1 + z)/2$ induces a bounded composition operator on $\ell^1(U)$. Generalize.

1.5   Notes

The result we are calling Littlewood's Subordination Principle is in Littlewood's paper [Ltw '25]. Actually it is part of the proof of a more general result: Littlewood showed that for any holomorphic self-map φ of U with $\varphi(0) = 0$, and for any $0 < p < \infty$,

$$\int_{-\pi}^{\pi} |f(\varphi(re^{i\theta}))|^p \, d\theta \leq \int_{-\pi}^{\pi} |f(re^{i\theta})|^p \, d\theta$$

for every $0 \leq r < 1$. The case $p = 2$ of this result follows upon applying our version of Littlewood's Subordination Principle to the dilated functions $f_r(z) = f(rz)$. Having proved the $p = 2$ case, Littlewood then reduced the general case to this one by a now-standard technique of "dividing out zeros" and taking an appropriate power of f.

This more general version of Littlewood's result shows that composition operators act continuously on each of the spaces H^p, defined for $0 < p < \infty$ as the collection of holomorphic functions on U for which

$$\|f\|_p^p = \sup_{0 \leq r < 1} \frac{1}{2\pi} \int_{-\pi}^{\pi} |f(re^{i\theta})|^p \, d\theta < \infty.$$

If $1 \leq p < \infty$ then $\|\cdot\|_p$ is a norm that makes H^p into a Banach space, and if $0 < p < 1$ then $\|f\|_p^p$ is a subadditive functional that induces a metric in which H^p is complete (see [Drn '70] and [Rdn '87, Chapter 17]) for more details).

The fact that composition operators preserve each of the spaces H^p can also be derived from the fact that the membership in the space is equivalent to the assertion that the subharmonic function $|f|^p$ has a harmonic majorant u on U. If this is the case, then $u \circ \varphi$ is a harmonic majorant for $|f \circ \varphi|^p$, which, therefore, also belongs to H^p. The boundedness of C_φ can also be derived from this argument by taking u to be the *least* harmonic majorant of f, in which case $\|f\| = u(0)^{1/p}$. Then $\|f \circ \varphi\| \leq u(\varphi(0))^{1/p}$, whereupon Harnack's inequality yields the analogue for general p of the bound we found for the norm of C_φ acting on H^2 (see [Drn '70, Th. 1.7, pp. 10–11] for more details).

Regarding Problem 10, Al'par [Alp '60] has shown that among the automorphisms of U, *only the rotations* take $\ell^1(U)$ into itself. Versions of this result for spaces lying properly between $\ell^1(U)$ and H^2 have been obtained by Halasz [Hlz '67].

Problems 4 through 11 show that composition operators can be studied in settings other than H^2. Arguably the most interesting alternate setting is $H^2(\mathbf{B})$, the H^2-space of the unit ball $\mathbf{B}$ of $\mathbf{C}^n$, as defined in [Rdn '80]. Most of the topics discussed in this book can be interpreted in this new setting, where they often lead to fascinating and difficult problems. For example: Does every holomorphic self-map of $\mathbf{B}$ induce a bounded composition operator on $H^2(\mathbf{B})$? The answer is *no*, and it is an interesting and non-trivial problem to decide which of these maps induce bounded operators. For a taste of what can happen when one studies boundedness and compactness of composition operators in several complex variables, see the last section of [MlS '86], and for a survey of recent work in the area, see [Wgn '90].

2
Compactness: Introduction

Having established that every composition operator is bounded on H^2, we turn to the most natural follow-up question:

Which composition operators are compact?

The property of "boundedness" for composition operators means that each one takes bounded subsets of H^2 to bounded subsets. The question above asks us to specify precisely how much the inducing map φ has to compress the unit disc into itself in order to insure that the operator C_φ compresses bounded subsets of H^2 into *relatively compact* ones.

This chapter will provide the background and intuition required to appreciate the problem. Its solution, at least for operators induced by univalent maps, will occupy the next chapter, while the one following that will treat the Julia-Carathéodory Theorem, which has been waiting for more than sixty years to give geometric meaning to our compactness criterion.

2.1 Compact Operators

A linear operator T on a Hilbert space $\mathcal{H}$ is said to be *compact* if it maps every bounded set into a relatively compact one (one whose closure in $\mathcal{H}$ is compact). It is not necessary to check *every* bounded set here; since translation and multiplication by a scalar are homeomorphisms of $\mathcal{H}$, it is enough to test only the unit ball.

By the Heine-Borel Theorem, every linear transformation on a finite dimensional Hilbert space is compact. Similarly, on an infinite dimensional

Hilbert space, every bounded operator with finite dimensional range is compact. Our first observation is that the compact operators are precisely those that can be approximated by such *finite rank operators*.

Finite Rank Approximation Theorem. *Suppose T is a bounded linear operator on a Hilbert space $\mathcal{H}$. Then T is compact if and only if there is a sequence $\{F_n\}$ of finite rank bounded operators such that $\|T - F_n\| \to 0$.*

Proof. Suppose first that T is compact on $\mathcal{H}$. Let $\{e_n\}$ be an orthonormal basis for $\mathcal{H}$ and consider the projection operators

$$P_n f = \sum_{k=0}^{n} <f, e_k> e_k \qquad (f \in \mathcal{H}).$$

where "$<,>$" denotes the inner product in $\mathcal{H}$. Clearly each operator P_n is a contraction on $\mathcal{H}$, and $\|P_n f - f\| \to 0$ for each $f \in \mathcal{H}$.

Let $\mathcal{B}$ denote the unit ball of $\mathcal{H}$ (open or closed, it doesn't matter). We are assuming that $T(\mathcal{B})$ is relatively compact in $\mathcal{H}$. Here is an easily proved, yet absolutely fundamental fact about sequences of transformations on a metric space:

> *Pointwise convergence plus equicontinuity implies uniform convergence on compact subsets.*

You encounter this principle, for example, when proving the Arzela-Ascoli Theorem (see for example [Rdn '76, Theorem 7.25], or [Frd '82, Theorem 3.6.4]). We have just observed that $P_n \to I$ (the identity map) pointwise on $\mathcal{H}$. Since the operators P_n are all contractions, the whole family is equicontinuous on bounded sets, so applying the equicontinuity principle stated above to the closure of $T(\mathcal{B})$, which we are assuming is compact, we see that $P_n \to I$ uniformly on $T(\mathcal{B})$. In other words, $P_n T \to T$ uniformly on $\mathcal{B}$, which just means that $\|P_n T - T\| \to 0$. Since $P_n T$ is a bounded finite rank operator for each n, this establishes the desired approximation.

Conversely, suppose some sequence $\{F_n\}$ of bounded finite rank operators converges in operator norm to T. We need to show that $T(\mathcal{B})$ is relatively compact in $\mathcal{H}$. Another result from metric space theory makes short work of this one ([Frd '82, Theorem 3.5.6]).

> *A subset $\mathcal{K}$ of a metric space X is relatively compact if and only if for every $\varepsilon > 0$ there is a finite set of points $\mathcal{N}_\varepsilon \subset X$ such that each point of $\mathcal{K}$ lies at most ε distant from $\mathcal{N}_\varepsilon$.*

The set $\mathcal{N}_\varepsilon$ is often called an "ε-net" for $\mathcal{K}$, and sets having an ε-net for every $\varepsilon > 0$ are often called "totally bounded." So a set is totally bounded if and only if it is relatively compact. If $\mathcal{K}$ is relatively compact, you get $\mathcal{N}_\varepsilon$ by covering the closure of $\mathcal{K}$ by open ε- balls, extracting a finite subcovering, and choosing as $\mathcal{N}_\varepsilon$ the centers. In the other direction, if you have $\mathcal{N}_\varepsilon$, you can use it (admittedly, with a little care) to produce, for any open covering of the closure of $\mathcal{K}$, a finite covering subordinate to the original one, from which follows the compactness of that closure (see, for example, [Nwm '61, Ch. II, Thm. 15.3], or [Frd '82, Theorem 3.5.6], for the details).

Returning to Hilbert space, let $\varepsilon > 0$ be given, and fix a value of n so that $\|F_n - T\| < \varepsilon/2$. Let $\mathcal{N}$ be an $\varepsilon/2$-net for $F_n(\mathcal{B})$. Then it is easy to check that $\mathcal{N}$ is an ε-net for $T(\mathcal{B})$. It follows from the characterization above that $T(\mathcal{B})$ is relatively compact. $\qquad\square$

This result suggests that the compact operators on Hilbert space should have much in common with operators on a finite dimensional space. In later chapters we will see more evidence of this principle. Right now it's time to use what we have learned to discover some interesting classes of compact composition operators.

2.2   First Class of Examples

The most drastic way φ can compress the unit disc is to take it to a point, in which case the resulting composition operator has one dimensional range (the space of constant functions), and is therefore compact. The next result shows that this compactness persists if we merely assume that $\varphi(U)$ is relatively compact in U (the rule-of-thumb here is that compact sets act like "large points.").

First Compactness Theorem. If $\|\varphi\|_\infty < 1$ then C_φ is a compact operator on H^2.

Proof. For each positive integer n define the operator

$$T_n f = \sum_{k=0}^{n} \hat{f}(k)\varphi^k \qquad (f \in H^2).$$

Thus T_n maps H^2 onto the linear span of the first n powers of φ. By our comparison of H^2 and H^∞ norms (§1.2), T_n is therefore a bounded, finite rank operator on H^2 (you can easily check that its norm is $\leq \sqrt{n+1}$). We claim that $\|C_\varphi - T_n\| \to 0$. This follows from the calculation below:

$$\|(C_\varphi - T_n)f\| = \left\| \sum_{k=n+1}^{\infty} \hat{f}(k)\varphi^k \right\|$$

$$\leq \sum_{k=n+1}^{\infty} |\hat{f}(k)| \, \|\varphi^k\|$$

$$\leq \sum_{k=n+1}^{\infty} |\hat{f}(k)|\, \|\varphi\|_\infty^k$$

$$\leq \left(\sum_{k=n+1}^{\infty} |\hat{f}(k)|^2 \right)^{\frac{1}{2}} \left(\sum_{k=n+1}^{\infty} \|\varphi\|_\infty^{2k} \right)^{\frac{1}{2}}$$

$$\leq \frac{\|\varphi\|_\infty^{n+1}}{\sqrt{1 - \|\varphi\|_\infty^2}}\, \|f\|,$$

where we have used respectively: the triangle inequality, the comparison of H^2 and H^∞ norms, the Cauchy-Schwarz inequality, and finally to sum the geometric series, the hypothesis that $\|\varphi\|_\infty < 1$. Thus

$$\|C_\varphi - T_n\| \leq \frac{\|\varphi\|_\infty^{n+1}}{\sqrt{1 - \|\varphi\|_\infty^2}} \to 0$$

as $n \to \infty$. This exhibits C_φ as an operator norm limit of finite rank operators, so it is compact on H^2.   □

This result shows that H^2 supports a lot of compact composition operators. The situation is, in fact, quite subtle; a small refinement of the proof shows that there are many others. In order to state the improved result without distracting complications, we need a boundary version of the integral representation of the H^2 norm developed in §1.2.

2.3   A Better Compactness Theorem

The H^2 norm revisited. Recall from Chapter 1 the proof that the standard conformal automorphisms α_p induce bounded composition operators on H^2. We first considered holomorphic polynomials f, for which the Proposition of §1.2 could be improved to:

$$\|f\|^2 = \frac{1}{2\pi} \int_{-\pi}^{\pi} |f(e^{i\theta})|^2 d\theta. \tag{1}$$

Clearly, this improvement carries over to more general functions, for example those that are continuous on the closed unit disc and holomorphic on the interior. Since the polynomials are dense in H^2 it seems reasonable that some form of this boundary representation of the norm should carry over to all of H^2. This is, in fact, the case. If $f \in H^2$, then because the coefficients are square-summable, the Fourier series $\sum_{n=0}^{\infty} \hat{f}(n)e^{in\theta}$ converges in L^2 to some $f^* \in L^2$, so clearly equation (1) holds with $f(e^{i\theta})$ replaced on the right by $f^*(e^{i\theta})$. What makes this formula really useful in the study of H^2 is something much deeper.

The Radial Limit Theorem. *Suppose $f(z) = \sum_{n=0}^{\infty} \hat{f}(n)z^n$ is a function in H^2, and f^* is the function in L^2 with Fourier series $\sum_0^{\infty} \hat{f}(n)e^{in\theta}$. Then*

$$\lim_{r \to 1-} f(re^{i\theta}) = f^*(e^{i\theta})$$

for almost every $e^{i\theta} \in \partial U$, and the H^2 norm of f is the L^2 norm of f^*.

The deep part of this theorem is the existence and identification of the radial limit function f^*. For a proof, see [Rdn '87, §17.11] or [Drn '70, Chapter 2]. Up to now we have tiptoed around this result. For example, it could have simplified the proof of boundedness for the operators C_{α_p}, making it unnecessary to first consider the special case of polynomials. From now on we will use this boundary form of the H^2 norm whenever it is convenient, always writing $f(e^{i\theta})$ instead of $f^*(e^{i\theta})$ for the radial limit.

However, this degree of generality is rarely needed to appreciate what is going on. Almost always, more elementary forms of the Radial Limit Theorem capture the spirit of the arguments. For example, in this chapter we will need the result only for holomorphic self-maps φ of the disc, so the bounded case suffices (see [Rdn '87, §11.32]). In fact, our results retain almost all their flavor even if φ is assumed smooth enough to extend continuously to the closed disc.

The better compactness theorem. Very early in the proof of our "First Compactness Theorem" we used the fact that the supremum norm dominates the H^2 norm. If we had waited a while, the crucial calculation would have looked like this:

$$\|(C_\varphi - T_n)f\| \leq \sum_{k=n+1}^{\infty} |\hat{f}(k)| \, \|\varphi^k\|$$

$$\leq \left(\sum_{k=n+1}^{\infty} |\hat{f}(k)|^2 \right)^{\frac{1}{2}} \left(\sum_{k=n+1}^{\infty} \|\varphi^k\|^2 \right)^{\frac{1}{2}},$$

which implies that

$$\|C_\varphi - T_n\| \leq \left(\sum_{k=n+1}^{\infty} \|\varphi^k\|^2 \right)^{\frac{1}{2}},$$

and, as before, this implies the compactness of C_φ provided that

$$\sum_{n=0}^{\infty} \|\varphi^n\|^2 < \infty. \tag{2}$$

Condition (2) can in turn be rewritten as follows, where we use the boundary form of the H^2 norm discussed in the last section.

$$\infty > \sum_{n=0}^{\infty} \frac{1}{2\pi} \int_{-\pi}^{\pi} |\varphi(e^{i\theta})|^{2n} d\theta$$

$$= \frac{1}{2\pi} \int_{-\pi}^{\pi} \sum_{n=0}^{\infty} |\varphi(e^{i\theta})|^{2n} d\theta$$

$$= \frac{1}{2\pi} \int_{-\pi}^{\pi} \frac{1}{1 - |\varphi(e^{i\theta})|^2} d\theta,$$

where the interchange of integration and summation is justified by positivity, and summation of the geometric series is already justified by its convergence (second line). We summarize our accomplishment in the following theorem, whose title will be explained shortly.

The Hilbert-Schmidt Theorem for composition operators. *If*

$$\int_{-\pi}^{\pi} \frac{1}{1 - |\varphi(e^{i\theta})|^2} d\theta < \infty, \tag{3}$$

then C_φ is compact on H^2.

Remark. The heart of our proof showed the integral condition (3) to be equivalent to (2), which can be rewritten $\sum \|C_\varphi(z^n)\|^2 < \infty$. More generally, an operator T on a Hilbert space $\mathcal{H}$ is called a *Hilbert-Schmidt operator* if, for some orthonormal basis $\{e_n\}$ of $\mathcal{H}$,

$$\sum_{n=0}^{\infty} \|Te_n\|^2 < \infty$$

(it is not difficult to show that if this condition holds for one orthonormal basis, then it holds for *all* of them [DdS '63, §XI.6, Lemma 2]). The argument that deduced the compactness of C_φ from (2) works in general, and shows:

Every Hilbert-Schmidt operator is compact.

The title of the Theorem above comes from the fact that its proof shows C_φ to be a Hilbert-Schmidt operator whenever φ satisfies (3). The fact that the Hilbert-Schmidt condition (2) does not depend on the particular choice of orthonormal basis, shows that our proof actually *characterizes* the Hilbert-Schmidt composition operators as the ones for which φ satisfies condition (3).

In the last section we showed that C_φ is compact whenever $\|\varphi\|_\infty < 1$. Our Hilbert Schmidt Theorem allows for a significant improvement.

The Polygonal Compactness Theorem. *If φ maps the unit disc into a polygon inscribed in the unit circle, then C_φ is compact on H^2.*

The proof will show that C_φ is actually a Hilbert-Schmidt operator. The major step involves proving the result for an important class of examples.

Lens Maps. For $0 < \alpha < 1$ define φ_α to be the holomorphic self-map of U that you get by using the Möbius transformation

$$\sigma(z) = \frac{1+z}{1-z} \qquad (4)$$

to take U onto the right half-plane, then employing the α-th power to squeeze the half-plane onto the sector $\{|\arg w| < \alpha\pi/2\}$, and completing the task by mapping back to U via σ^{-1}. The result is:

$$\varphi_\alpha(z) = \frac{\sigma(z)^\alpha - 1}{\sigma(z)^\alpha + 1}. \qquad (5)$$

Because φ_α takes the unit disc onto the lens-shaped region L_α shown below, we call it a "lens map." Our first result asserts that each lens map induces

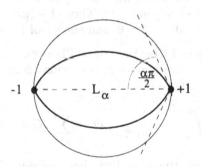

FIGURE 2.1. The lens L_α

a Hilbert-Schmidt operator on H^2.

Lemma. *Each lens map φ_α $(0 < \alpha < 1)$ satisfies the Hilbert-Schmidt condition (3).*

Proof. For convenience we write φ instead of φ_α. Since φ fixes the points ± 1 and sends every other point of ∂U into U, it is enough to examine the integrability of $(1 - |\varphi(e^{i\theta})|^2)^{-1}$ over small arcs centered at ± 1, and by symmetry it is enough to consider just one of these points, say $+1$. To study the behavior of φ near this point, observe that

$$1 - \varphi(z) = \frac{2}{\sigma(z)^\alpha + 1}.$$

A little calculation shows that $\sigma(e^{i\theta}) = i \cot(\theta/2)$, so for $|\theta| < \pi/2$,

$$|\sigma(e^{i\theta})| = |\cot(\theta/2)| \leq \frac{2}{|\theta|},$$

whereupon

$$|1 - \varphi(e^{i\theta})| \geq \frac{2}{|\sigma(e^{i\theta})|^\alpha + 1}$$

$$\geq \text{ const. } |\theta|^\alpha.$$

Since $0 < \alpha < 1$ this estimate implies that the function $|1 - \varphi(e^{i\theta})|^{-1}$ is integrable over the interval $[-\pi/2, \pi/2]$.

Now, each point $\varphi(e^{i\theta})$ lies between the real axis and a line through the point $+1$ that makes an angle $\alpha\pi/2$ with that axis. An exercise involving the law of cosines shows that because of this,

$$1 - |\varphi(e^{i\theta})|^2 \geq \text{const. } |1 - \varphi(e^{i\theta})|$$

for all θ near 0 (i.e. the distance from $\varphi(e^{i\theta})$ to the unit circle is about the same as its distance to the point $+1$). Thus $(1 - |\varphi(e^{i\theta})|^2)^{-1}$ is integrable in an interval centered about $\theta = 0$, and the proof is complete.   □

Proof of the Polygonal Compactness Theorem. Recall that the proof of our Hilbert-Schmidt theorem for composition operators showed that for any holomorphic self-map φ of U,

$$\frac{1}{2\pi} \int_{-\pi}^{\pi} \frac{1}{1 - |\varphi(e^{i\theta})|^2} \, d\theta = \sum_{n=0}^{\infty} \|C_\varphi(z^n)\|^2, \tag{6}$$

where we now allow for the possibility that one side of the equation (and therefore both sides) might be infinite.

Suppose, to begin our proof, that φ maps the unit disc into one of the lenses L_α. Then $\psi = \varphi_\alpha^{-1} \circ \varphi$ is a holomorphic self-map of U, and $\varphi = \varphi_\alpha \circ \psi$. Thus $C_\varphi = C_\psi C_{\varphi_\alpha}$, so

$$\|C_\varphi(z^n)\| \leq \|C_\psi\| \|C_{\varphi_\alpha}(z^n)\|$$

for each non-negative integer n. Our Lemma about lens maps, along with (6) above shows that

$$\frac{1}{2\pi} \int_{-\pi}^{\pi} \frac{1}{1 - |\varphi(e^{i\theta})|^2} \, d\theta \leq \|C_\psi\|^2 \sum_{n=0}^{\infty} \|C_{\varphi_\alpha}(z^n)\|^2 < \infty. \tag{7}$$

This shows that anything that maps the unit disc into a lens induces a Hilbert-Schmidt operator.

Now for the general case. The factorization argument above shows that it is enough to consider maps φ that take the unit disc *conformally onto* polygons inscribed in the unit circle. Each such φ extends to a homeomorphism from the closed disc onto the closure of the polygon (you can justify this by either quoting the elementary theory of the Schwarz-Christoffel transformation [ChB '90, Chapter 10] or the Carathéodory extension theorem, as developed in [Rdn '87, Theorem 14.18]). Consider a vertex of the polygon, which, without loss of generality, we may assume to be the point $+1$. We may also assume this point is fixed by φ. Thus the map $\chi = (1+\varphi)/2$ fixes $+1$ and takes the disc into a lens L_α for some α sufficiently close to 1, so by the work of the last paragraph the function $(1 - |\chi(e^{i\theta})|^2)^{-1}$ is integrable over the unit circle. Now as $\theta \to 0$, both $\varphi(e^{i\theta})$ and $\chi(e^{i\theta})$ approach $+1$, while staying inside L_α (i.e. they approach $+1$ "non-tangentially"), so we have for all sufficiently small θ:

$$1 - |\chi(e^{i\theta})|^2 \approx |1 - \chi(e^{i\theta})| = |\frac{1 - \varphi(e^{i\theta})}{2}| \approx \frac{1 - |\varphi(e^{i\theta})|^2}{2}.$$

Thus, the reciprocal of the function on the right is integrable over an interval symmetric about $\theta = 0$.

The function $(1 - |\varphi(e^{i\theta})|^2)^{-1}$ is therefore integrable over an interval centered about the preimage of each vertex of the polygon, so it is therefore integrable over the whole unit circle. Our Hilbert-Schmidt Theorem now shows that C_φ is compact on H^2. $\qquad\square$

In the next chapter we will see that compactness survives if the corners of the polygon are rounded "just a little." Right now we intend to show that compactness can be defeated by rounding these corners "too much."

2.4   Compactness and Weak Convergence

When studying compactness in metric spaces it often helps to express everything in terms of sequential convergence. The same holds for the study of compact operators. The definition of compactness for Hilbert space operators can be rephrased to read something like "compact operators are the ones that take weakly convergent sequences into norm convergent ones." Here is the version best suited to our purposes.

Weak Convergence Theorem. *For φ a holomorphic self-map of U, the following statements are equivalent.*

(a) *C_φ is a compact operator on H^2.*

(b) *If $\{f_n\}$ is a sequence that is bounded in H^2 and converges to zero uniformly on compact subsets of U, then $\|C_\varphi f_n\| \to 0$.*

Proof. The key to this proof is the fundamental growth condition (2) of §1.1, which asserts that H^2 convergence implies pointwise convergence on U, and that bounded subsets of H^2 are, as classes of functions, uniformly bounded on compact subsets of U. Let $\mathcal{B}$ denote the closed unit ball in H^2.

(a) $\Rightarrow$(b): We are assuming that C_φ is a compact operator, i.e., that $C_\varphi(\mathcal{B})$ is a relatively compact subset of H^2. We are further given a sequence $\{f_n\}$ that lies in $M\mathcal{B}$ (the ball of radius M), and converges to zero uniformly on compact subsets of U (in symbols $f_n \xrightarrow{\kappa} 0$). We have to show that $\|C_\varphi f_n\| \to 0$, and for this it suffices to show that the zero-function is the unique limit point of the sequence $\{C_\varphi f_n\}$ (for the norm topology). But $C_\varphi f_n \xrightarrow{\kappa} 0$, and since H^2 convergence implies pointwise convergence, zero is the only possible limit point. By the compactness of C_φ, the set $\{C_\varphi f_n\}$ is relatively compact, so there must be a limit point, and we are done.

(b) $\Rightarrow$(a): Suppose $\{f_n\}$ is a sequence of functions in $\mathcal{B}$. We wish to show that the image sequence $\{C_\varphi f_n\}$ has a convergent subsequence. Because the functions in $\mathcal{B}$ are bounded uniformly on compact subsets of U, Montel's Theorem [Rdn '87, Theorem 14.6, page 282] picks out a subsequence $\{g_k = f_{n_k}\}$ that converges uniformly on compact subsets of U to a holomorphic function g. We claim that $g \in H^2$. Indeed, for each $0 < r < 1$,

$$\frac{1}{2\pi} \int_{-\pi}^{\pi} |g(re^{i\theta})|^2 d\theta = \lim_{k \to \infty} \frac{1}{2\pi} \int_{-\pi}^{\pi} |g_k(re^{i\theta})|^2 d\theta \leq \sup_k \|g_k\|^2 \leq 1.$$

This inequality, along with the integral characterization of H^2 and its norm (§1.2), shows that $g \in H^2$, and in fact $\|g\| \leq 1$.

Thus the sequence $\{g_k - g\}$ is bounded in H^2, and $g_k - g \xrightarrow{\kappa} 0$. Therefore hypothesis (b) insures that $\|C_\varphi(g_k - g)\| \to 0$, as desired. $\qquad\square$

For a more general version of this result, see Exercises 8 and 9 below. In order to appreciate its power, note how the result makes short work of the First Compactness Theorem of §2.2.

2.5 Non-Compact Composition Operators

We use the "sequential" characterization of compactness to show that C_φ can fail to be compact if $\varphi(e^{i\theta})$ approaches the boundary of U either too quickly or too often. Our first example shows that C_φ can fail to be compact even if $|\varphi(e^{i\theta})| = 1$ at a single point $e^{i\theta} \in \partial U$.

Example (The values of φ approach the boundary too quickly). For $0 < \lambda < 1$ let $\varphi(z) = \lambda z + (1 - \lambda)$. Then C_φ is not compact on H^2.

Proof. For each fixed $0 < r < 1$ define

$$f_r(z) = \frac{\sqrt{1 - r^2}}{1 - rz} \qquad (z \in U).$$

From the original definition of the H^2 norm in terms of power series coefficients, each of these functions has norm 1, and it is clear that $f_r \to 0$
uniformly on compact subsets of U as $r \to 1-$. An exercise involving geometric series shows that

$$\|f_r \circ \varphi\|^2 = \frac{1+r}{1+r(2\lambda - 1)} \qquad (0 < r < 1),$$

so $\|f_r\| \to \lambda^{-1/2} \neq 0$ as $r \to 1-$. By the work of the last section, this shows
that C_φ is not compact.   □

Intuitively, if a map induces a non-compact operator, then any map
whose values approach the unit circle "faster" should also induce a non-
compact operator. The theorem below formalizes this idea, and allows us
to turn results for specific classes of maps like the ones above into general
compactness theorems.

Comparison Principle for Compactness. *Suppose φ and ψ are holomorphic self-maps of U, with φ univalent and $\psi(U) \subset \varphi(U)$. If C_φ is
compact on H^2, then so is C_ψ.*

Proof. We use an argument similar to the one that occurred in the proof
of the Polygonal Compactness Theorem. Because φ is univalent and its
range contains that of ψ, we can form the map $\chi = \varphi^{-1} \circ \psi$, which takes U
holomorphically into itself. Thus $\psi = \varphi \circ \chi$, so on the operator level, $C_\psi =
C_\chi C_\varphi$. The desired result now follows from a very simple, but enormously
useful fact about compact operators:

> If S and T are operators on a Hilbert space $\mathcal{H}$, with S bounded
> and T compact, then both ST and TS are compact.

The proof follows immediately from the definition of compact operator, and
the fact that bounded operators preserve both boundedness and relative
compactness of subsets of $\mathcal{H}$. We leave the details to the reader.   □

We can use our new Comparison Principle to generalize the class of
examples that led off this section.

Corollary. *Suppose φ is a univalent self-map of U, and that $\varphi(U)$ contains
a disc in U that is tangent to the unit circle. Then C_φ is not compact.*

Proof. We may suppose, without loss of generality, that the disc (call it
Δ) is tangent to the unit circle at $+1$. Therefore if λ is the radius of Δ, we
have $0 < \lambda < 1$ and $\Delta = \lambda U + (1 - \lambda)$. Thus Δ is the image of the U under
the map $\psi(z) = \lambda z + (1 - \lambda)$, which by the first result of this section is not
compact. By the Comparison Principle, C_φ is not compact either.   □

Remarks. (a) In the Comparison Principle we cannot do without the univalence of φ. Indeed, there exists a map that takes U *onto* itself in no more than two-to-one fashion, but nonetheless induces a compact composition operator (see §3.7, Exercise 7).

(b) In Chapter 4 we will show that non-compactness persists if $\psi(U)$ contains a domain whose boundary approaches the unit circle "as smoothly as the curve $y = x^\alpha$ approaches the real axis $(1 < \alpha \le 2)$." The corollary above deals with the case $\alpha = 2$, while the Polygonal Compactness Theorem shows that the result fails for $\alpha = 1$.

(c) *Subordination.* In the proofs of both the Polygonal Compactness Theorem and the above Comparison Principle we used the fact that:

> If φ and ψ are holomorphic self-maps of U with φ univalent and $\psi(U) \subset \varphi(U)$, then $\psi = \varphi \circ \chi$ where χ is a holomorphic self-map of U .

More generally, if f and g are any two holomorphic functions, with $f = g \circ \chi$ where χ is a holomorphic self-map of U, we say f is *subordinate* to g (in the usual definition of subordination it is also required that $\chi(0) = 0$, but here we ignore this). The idea of subordination has also come up in Exercises 2 and 3 of Chapter 1.

The above results assert that a composition operator cannot be compact if the values of its inducing map approach the unit circle too quickly, even if this only happens at a single point. Here is an apparently different way to defeat compactness (the ominous adverb "apparently" will be explained in the next two chapters).

Proposition (The values of φ approach the boundary too often). *Suppose φ is a holomorphic self-map of U for which the set*

$$E(\varphi) = \{\theta \in [-\pi, \pi] : |\varphi(e^{i\theta})| = 1\}$$

has positive Lebesgue measure. Then C_φ is not compact on H^2.

Proof. Set $E = E(\varphi)$. Clearly each monomial z^n $(n \ge 0)$ belongs to the unit ball of H^2, and the whole sequence tends to zero uniformly on compact subsets of U. On the other hand,

$$\|C_\varphi(z^n)\|^2 = \frac{1}{2\pi} \int_{-\pi}^{\pi} |\varphi(e^{i\theta})|^{2n} d\theta$$

$$\ge \frac{1}{2\pi} \int_E |\varphi(e^{i\theta})|^{2n} d\theta$$

$$\ge \frac{1}{2\pi} |E| > 0$$

where $|E|$ denotes the Lebesgue measure of E. Thus the sequence $\{C_\varphi(z^n)\}$ does not tend to zero in norm, so C_φ is not a compact operator. $\square$

Summary. We have seen that C_φ is compact if $\varphi(z)$ stays inside an inscribed polygon, but that it is *not* compact whenever $\varphi(z)$ approaches to the unit circle "too often," in the sense that $|\varphi(e^{i\theta})| = 1$ for θ in a set of positive measure, or "too closely," as is the case for a univalent inducing map whose image contains a disc that is tangent to the unit circle. These results all suggest that

> C_φ is compact if and only if $\varphi(z)$ does not get too close to the unit circle too often.

In the next two chapters we will come to understand precisely what this means for *univalent maps* φ, and in Chapter 10 (which is independent of almost all the intervening material) we will see what it means in general.

2.6 Exercises

The first three exercises below explore the concepts of boundedness and compactness for the simplest Hilbert space operators: the *diagonal* ones.

1. For each bounded complex sequence $\Lambda = \{\lambda_n\}$ define the *diagonal operator* D_Λ on the sequence space ℓ^2 by:

$$D_\Lambda x = \{\lambda_n \xi_n\} \qquad (x = \{\xi_n\} \in \ell^2).$$

   Show that D_Λ is a bounded operator on ℓ^2, compute its norm, and show that it is compact if and only if $\lambda_n \to 0$.

2. Show that D_Λ is a Hilbert-Schmidt operator on ℓ^2 if and only if $\Lambda \in \ell^2$.

3. Show that the distance from a bounded diagonal operator D_Λ to the subspace of compact operators is $\limsup_n |\lambda_n|$.

4. *A non-compact composition operator with compact square.* For the map $\varphi(z) = (1 - z)/2$, show that C_φ^2 is compact, but that C_φ itself is not. Can this phenomenon happen for diagonal operators?

5. Characterize the Hilbert-Schmidt composition operators on the Bergman space A^2 (see Problem 4 of §1.4).

6. Characterize the Hilbert-Schmidt composition operators on the Dirichlet space (see Problems 6 through 9 of §1.4).

7. *A metric for uniform convergence on compact sets.* For $f \in H(U)$ and n a positive integer, define $\|f\|_n$ to be the maximum of $|f(z)|$ for $|z| \leq 1 - \frac{1}{n}$. Let

$$\nu_n(f) = \frac{\|f\|_n}{1 + \|f\|_n} \quad \text{and} \quad \nu(f) = \sum_{n=1}^{\infty} \frac{1}{2^n} \nu_n(f).$$

Show that:

(a) A sequence $\{f_k\}$ in $H(U)$ converges to zero uniformly on compact subsets of U if and only $\nu(f_n) \to 0$.

(b) The function $d(f, g) \overset{\text{def}}{=} \nu(f - g)$ defined for $f, g \in H(U)$ is a metric on $H(U)$ in which convergence is equivalent to uniform convergence on compact subsets of U.

Suggestion: The key to showing that d is a metric lies in showing that ν is *subadditive*, i.e., that $\nu(f+g) \leq \nu(f)+\nu(g)$ for each $f, g \in H(U)$. For this it is enough to show that each ν_n has this property, and for this you only have to show that the function $x/(1+x)$ is subadditive on the positive real axis.

8. A sequence $\{x_n\}$ in a Hilbert space $\mathcal{H}$ is said to converge *weakly* to $x \in \mathcal{H}$ if $< x_n, y > \to < x, y >$ for every $y \in \mathcal{H}$.

(a) Show that every weakly convergent sequence is bounded.

(b) Show that a bounded operator T on Hilbert space is compact if and only if $\|Tx_n\| \to 0$ whenever $x_n \to 0$ weakly.

9. (a) Show that a sequence in H^2 is weakly convergent to 0 if and only if it is bounded and converges to 0 uniformly on compact subsets of U.

(b) Use part (a) above with Exercise 8 to give another proof of the Weak Convergence Theorem.

10. H^∞ is a Banach space in the norm $\| \cdot \|_\infty$. Show that a composition operator C_φ is compact on H^∞ if and only if $\|\varphi\|_\infty < 1$. (Just as in Hilbert space, an operator on a Banach space is *compact* if it takes the unit ball to a relatively compact set.)

11. Prove a similar result for the space $\ell^1(U)$ considered in Problem 10 of §1.4.

2.7 Notes

The operator theoretic approach to subordination seems to have first appeared in the work of Ryff [Rff '66] and Nordgren [Ngn '68]. The study of compactness was initiated by H.J. Schwartz [Schw '69], who obtained the results of §2.2, and showed that the integrability of $(1 - |\varphi|)^{-1}$ over the unit circle implies compactness.

The connection with Hilbert-Schmidt operators and the fact that C_φ is Hilbert-Schmidt whenever $\varphi(U)$ lies in an inscribed polygon comes from [STa '73], where it is shown that such operators actually lie in every "Schatten p-class" (see the *Notes* for Chapter 10 for further comments and references on this topic). Examples of compact composition operators that are *not* Hilbert-Schmidt were constructed in [STa '73]; see Exercise 12 at the end of Chapter 3.

The compactness problem makes sense for the other H^p spaces, as defined in the *Notes* for Chapter 1, but leads to nothing new. Problem 10 above shows that the problem is not interesting for H^∞, while for $0 < p < \infty$ it is known that C_φ is compact on H^p if and only if it is compact on H^2 (see §6 of [STa '73]).

In a more fruitful direction, Banach space theorists have identified various subclasses of compact operators that generalize classes like the Hilbert-Schmidt operators on Hilbert space. Jarchow and Huniziger have done interesting work on the question of which composition operators on H^p (or between different H^p spaces) belong to such classes; see [Hzr '89], [HzJ '91], [Jrw '92], and [Jrw '93].

3
Compactness and Univalence

We are now ready to classify the univalent self-maps of U that induce compact composition operators on H^2. A fragment of operator theoretic folk-wisdom will help us guess the answer:

> If a "big-oh" condition describes a class of bounded operators, then the corresponding "little-oh" condition picks out the sub-class of compact operators.

(Problem 1 of §2.6 shows this principle in action). Unfortunately, no hint of any "big-oh" condition emerged from our proof of Littlewood's Theorem, so in this chapter the first order of business has to be:

> Find the "right" proof of boundedness for composition operators acting on H^2.

Given what we have already done, it should come as no surprise that everything will depend on getting the "right" representation of the H^2 norm.

3.1 The H^2 Norm via Area Integrals

At one time or another we have employed each of the following formulas for the norm of a function $f \in H^2$:

$$\|f\|^2 \;=\; \sum_{n=0}^{\infty} |\hat{f}(n)|^2 \qquad \text{(the definition)}$$

$$= \lim_{r \to 1-} \frac{1}{2\pi} \int_{-\pi}^{\pi} |f(re^{i\theta})|^2 d\theta$$

$$= \frac{1}{2\pi} \int_{-\pi}^{\pi} |f(e^{i\theta})|^2 d\theta,$$

where the second equation was derived in §1.2 and the third in §2.3. This section will contribute one more item to the list: a representation of the norm by an integral over the unit disc itself. In what follows we write dA for two dimensional Lebesgue measure, restricted to the unit disc, and normalized to have mass one ($dA = \frac{1}{\pi} dx dy$). We continue with the convention that the H^2 norm can act on any $f \in H(U)$, taking the value ∞ on those functions that do not belong to H^2.

Proposition (Area integral estimate for the H^2 norm). *For $f \in H(U)$,*

$$\frac{1}{2}\|f - f(0)\|^2 \leq \int_U |f'(z)|^2(1 - |z|^2) dA(z) \leq \|f - f(0)\|^2. \tag{1}$$

Proof. The idea is to write the integral in polar coordinates, use Fubini's Theorem, and see what emerges.

$$\int_U |f'(z)|^2(1 - |z|^2) \, dA(z) = 2\int_0^1 \left(\frac{1}{2\pi} \int_{-\pi}^{\pi} |f'(re^{i\theta})|^2 \, d\theta \right) (1 - r^2) \, r dr$$

$$= 2\int_0^1 \left(\sum_{n=1}^{\infty} n^2 |\hat{f}(n)|^2 r^{2n-2} \right) (1 - r^2) \, r dr$$

$$= 2\sum_{n=1}^{\infty} n^2 |\hat{f}(n)|^2 \int_0^1 r^{2n-2}(1 - r^2) \, r dr$$

$$= \sum_{n=1}^{\infty} \frac{n}{n+1} |\hat{f}(n)|^2.$$

Clearly the last quantity lies between $\|f - f(0)\|^2$ and half this quantity, as promised.   □

A representation of the norm that gives an identity rather than just a pair of inequalities can be proved at the expense of replacing the weight $1 - |z|^2$ by $-2\log|z|$ (see Exercise 1 below).

3.2   The Theorem

Littlewood's Theorem revisited. Armed with the area integral representation of the H^2 norm, we can now give the "right" proof of Littlewood's

Theorem, at least for univalent inducing maps with $\varphi(0) = 0$. For $f \in H^2$, we substitute $f \circ \varphi$ for f in the above equation, and use in succession the chain rule, the Schwarz Lemma [Rdn '87, Theorem 12.2], and the change of variable $w = \varphi(z)$:

$$\frac{1}{2}\|f \circ \varphi - f(0)\|^2 \leq \int_U |(f \circ \varphi)'(z)|^2 (1 - |z|^2) \, dA(z)$$

$$= \int_U |f'(\varphi(z))|^2 (1 - |z|^2) |\varphi'(z)|^2 \, dA(z)$$

$$\leq \int_U |f'(\varphi(z))|^2 (1 - |\varphi(z)|^2) |\varphi'(z)|^2 \, dA(z)$$

$$= \int_{\varphi(U)} |f'(w)|^2 (1 - |w|^2) \, dA(w)$$

$$\leq \|f - f(0)\|^2,$$

where the last inequality follows from the upper estimate of the area integral representation. By the definition of H^2 norm (in terms of series) the right-hand side of the inequality above is $\leq \|f\|^2$. Along with some final calculation involving the triangle inequality and the fact that $|f(0)| \leq \|f\|$, this shows that $\|C_\varphi f\| \leq 3\|f\|$, hence C_φ is bounded on H^2.

At first glance this method of proving boundedness for composition operators seems to lose a lot, since we actually know that C_φ is a *contraction* whenever $\varphi(0) = 0$, even if φ is not univalent. But the method provides what we need most: the "big-oh" condition that stands behind Littlewood's Theorem; it is nothing but *the Schwarz Lemma*, disguised in the form

$$1 - |z|^2 = O(1 - |\varphi(z)|^2) \quad \text{as} \quad |z| \to 1-,$$

where the "big-oh" constant is 1. According to the Folk Wisdom dispensed at the beginning of this chapter, the corresponding "little-oh" condition should tell a lot about compactness. This intuition is confirmed by the next result, which is the main result of this chapter.

The Univalent Compactness Theorem. *Suppose φ is a univalent self-map of U. Then C_φ is compact on H^2 if and only if*

$$\lim_{|z| \to 1-} \frac{1 - |\varphi(z)|}{1 - |z|} = \infty. \tag{2}$$

The proof of this result will occupy the rest of the chapter. The geometric interpretation of condition (2) will be provided in the next chapter by the Julia-Carathéodory Theorem.

3.3 Proof of Sufficiency

We are assuming that φ is univalent and satisfies condition (2). To show
that C_φ is compact on H^2 we employ the "sequential" characterization of
compactness given in §2.4. To this end, suppose $\{f_n\}$ is a sequence that is
bounded in H^2, and converges to zero uniformly on compact subsets of U.
Our goal is to show that $\|C_\varphi f_n\| \to 0$. Without loss of generality we may
assume that $\|f_n\| \le 1$ for all n.

Let $\varepsilon > 0$ be given. Then condition (2) guarantees a number $0 < r < 1$
so that

$$1 - |z|^2 \le \varepsilon(1 - |\varphi(z)|^2) \qquad \text{for } r < |z| < 1. \tag{3}$$

We fix this r for the remainder of the proof. According to the area integral
estimate (1) of the H^2 norm:

$$\frac{1}{2}\|C_\varphi f_n - f_n(\varphi(0))\|^2 \le \int_{rU} + \int_{U \setminus rU} |(f_n \circ \varphi)'(z)|^2(1 - |z|^2)\, dA(z).$$

Since $f_n \circ \varphi \to 0$ uniformly on compact subsets of U, the same is true of
the sequence of derivatives. Therefore, the first integral above converges to
zero, so in what follows we simply denote it by "$o(1)$." Our real concern
is the *second* integral, which we estimate by successively using inequality
(3), replacing the annulus by the whole disc, and changing variables as in
the last section. These steps yield the following string of inequalities:

$$\|C_\varphi f_n - f_n(\varphi(0))\|^2$$

$$\le \quad o(1) + \varepsilon \int_{U \setminus rU} |f_n(\varphi(z))\varphi'(z)|^2(1 - |\varphi(z)|^2)\, dA(z)$$

$$\le \quad o(1) + \varepsilon \int_U |f_n'(\varphi(z))|^2(1 - |\varphi(z)|^2)|\varphi'(z)|^2\, dA(z)$$

$$\le \quad o(1) + \varepsilon \int_U |f_n'(w)|^2(1 - |w|^2)\, dA(w)$$

$$\le \quad o(1) + 2\varepsilon\|f_n - f_n(0)\|^2 \qquad \text{[by (1) again]}$$

$$\le \quad o(1) + 2\varepsilon,$$

where in the last line we used the fact that $\|f_n - f_n(0)\| \le \|f_n\| \le 1$ for each
n. Since $f_n(\varphi(0)) \to 0$, the estimate above shows that $\limsup_n \|C_\varphi f_n\| \le$
2ε. Because ε was an arbitrary positive number, this shows $\|C_\varphi f_n\| \to 0$,
and completes the proof that C_φ is compact. □

The more subtle part of the theorem is the proof that condition (2) is *necessary* for compactness. There are several paths to this result, each of which requires a new idea. We choose one based on elementary operator theory. Later on, in solving the general compactness problem (Chapter 10), we will give another approach based on subharmonicity.

3.4   The Adjoint Operator

The scene now shifts to an abstract (separable) Hilbert space $\mathcal{H}$. Recall that the norm of each element $y \in \mathcal{H}$ can be expressed in terms of the inner product by

$$\|y\| = \sup_{x \in \mathcal{B}} |<x, y>|, \tag{4}$$

where $\mathcal{B}$ is the closed unit ball in $\mathcal{H}$, and the supremum is attained at the unit vector $x = y/\|y\|$ (assuming $y \neq 0$). In particular, the linear functional induced on $\mathcal{H}$ by y:

$$x \mapsto <x, y> \qquad (x \in \mathcal{H})$$

is a bounded linear functional on $\mathcal{H}$ of norm $\|y\|$. The Riesz Representation Theorem ([Rdn '87], Theorem 4.12) asserts that each bounded linear functional on $\mathcal{H}$ is induced in this way by some (necessarily unique) vector $y \in \mathcal{H}$.

Now if T is a bounded linear operator on $\mathcal{H}$, and $y \in \mathcal{H}$, then the linear functional

$$x \mapsto <Tx, y> \qquad (x \in \mathcal{H})$$

is bounded, so there is a unique vector in $\mathcal{H}$, which we denote by T^*y, that represents this functional in equation (4). The operator T^* so defined on $\mathcal{H}$ is called the *adjoint* of T; its definition can be summarized like this:

$$<x, T^*y> = <Tx, y> \qquad (x, y \in \mathcal{H}). \tag{5}$$

Clearly T^* is a linear transformation on $\mathcal{H}$, and (4) implies that $\|T^*\| = \|T\|$. It is also routine to check that

$$(T_1 + T_2)^* = T_1^* + T_2^* \qquad \text{and} \qquad (cT)^* = \bar{c}T^*.$$

where the symbol T, with or without subscripts, denotes a bounded linear operator on $\mathcal{H}$, and c is a complex number. In short, we have proved the following result, where $\mathcal{L}(\mathcal{H})$ denotes the space of bounded linear operators on $\mathcal{H}$, taken in the operator norm.

Lemma. *The map $T \to T^*$ is a conjugate-linear isometry on $\mathcal{L}(\mathcal{H})$.*

The adjoint of a finite rank operator. Here is an important example of the computation of the adjoint operator. Suppose T is a bounded operator *of rank one* on $\mathcal{H}$. This means that for some $x, y \in \mathcal{H}$,

$$Tz = < z, y > x \qquad (z \in \mathcal{H}).$$

Using equation (5) we easily compute for each $z, w \in \mathcal{H}$

$$
\begin{aligned}
< z, T^* w > \ &= \ < Tz, w > \\
&= \ < z, y >< x, w > \\
&= \ < z, \overline{< x, w >} y > \\
&= \ < z, < w, x > y >,
\end{aligned}
$$

from which it follows that $T^* w = < w, x > y$.

It is customary to write the one dimensional operator T as $x \otimes y$, so the result just proved can be succinctly rephrased as follows.

Lemma. *The adjoint of a rank one operator has rank one; in fact if $x, y \in \mathcal{H}$, then $(x \otimes y)^* = y \otimes x$.*

Since every finite rank operator is a sum of rank one operators, the Lemma and the linear nature of the adjoint operation yield this:

Corollary. *The adjoint of a finite rank bounded operator again has finite rank.*

All the results developed in this section now combine to show that the adjoint operation preserves compactness.

Proposition. *The adjoint of a compact operator is compact.*

Proof. Suppose T is a compact operator on $\mathcal{H}$. By the approximation theorem of §2.1 we know that there exists a sequence $\{F_n\}$ of bounded finite rank operators such that $\|T - F_n\| \to 0$. Since the adjoint operation is additive and isometric in the operator norm,

$$\lim_n \|T^* - F_n^*\| = \lim_n \|(T - F_n)^*\| = 0.$$

Since each of the operators F_n^* is of finite rank (and bounded), another appeal to the approximation theorem shows that T^* is compact.   □

Adjoint composition operators and reproducing kernels. Our second computation involves the adjoint of a composition operator. Although there is no good description of the adjoint that works for all composition operators on all H^2 functions, we can always compute its action on an important special family of functions in H^2: the *reproducing kernels*.

For each point $p \in U$, let

$$k_p(z) \overset{\text{def}}{=} \frac{1}{1 - \bar{p}z} = \sum_{n=0}^{\infty} \bar{p}^n z^n.$$

Clearly $k_p \in H^2$. It is called the *reproducing kernel* for the point p, and it gets the name from the fact that for each $f \in H^2$,

$$< f, k_p > = \sum_{n=0}^{\infty} \hat{f}(n) p^n = f(p). \tag{6}$$

Lemma. $C_\varphi^* k_p = k_{\varphi(p)}$ *for each* $p \in U$.

Proof. For each $f \in H^2$ we have

$$< f, C_\varphi^* k_p > = < C_\varphi f, k_p > = C_\varphi f(p) = f(\varphi(p)) = < f, k_{\varphi(p)} >$$

which proves the result. $\square$

3.5 Proof of Necessity

Equipped with our newly constructed adjoint machinery, we can finish the proof of the Univalent Compactness Theorem. It turns out that our method does not require univalence, so it yields a completely general necessary condition for compactness.

Theorem (Necessary condition for compactness). *Suppose* φ *is a holomorphic self-map of* U *and that* C_φ *is compact on* H^2. *Then*

$$\lim_{|z| \to 1-} \frac{1 - |\varphi(z)|}{1 - |z|} = \infty.$$

Proof. For each $p \in U$ let

$$f_p(z) = \frac{k_p}{\|k_p\|} = \frac{\sqrt{1 - |p|^2}}{1 - \bar{p}z} \, ,$$

the *normalized reproducing kernel* for p. (The impression of *deja vu* here is real; we previously encountered the functions f_p for p in the unit interval in the course of analyzing the first class of examples in §2.5.) We are going to show that

$$\|C_\varphi^* f_p\| \to 0 \quad \text{as} \quad |p| \to 1-. \tag{7}$$

This will finish the proof, since the last Lemma implies

$$\|C_\varphi^* f_p\|^2 = (1 - |p|^2)\, \|k_{\varphi(p)}\|^2 = \frac{1 - |p|^2}{1 - |\varphi(p)|^2}.$$

(One is tempted to try to deduce (7) from the Weak Convergence Theorem of §2.4, but that result was proved only for composition operators, *not* for their adjoints. However, readers experienced in Hilbert space arguments might wish to omit the argument below, and instead obtain (7) in a less *ad hoc* manner from Exercises 8 and 9 of §2.6.)

To prove (7), recall from §3.4 that the adjoint operator C_φ^* inherits the compactness of C_φ. Thus the collection of C_φ^* images of normalized reproducing kernels is a relatively compact subset of H^2, so every sequence of these images has a convergent subsequence. We need only show that the zero function is the only possible limit of such a subsequence.

So it all comes down to this: Suppose $|p_n| \to 1-$, and $C_\varphi^* f_{p_n} \to g$ in the H^2 norm. We'll be finished if we can show that $g = 0$. To see this, let h be any polynomial. Then the continuity of the inner product gives

$$
\begin{aligned}
< g, h > \quad &= \quad \lim_n < C_\varphi^* f_{p_n}, h > \\[2ex]
&= \quad \lim_n \sqrt{1 - |p_n|^2} < C_\varphi^* k_{p_n}, h > \\[2ex]
&= \quad \lim_n \sqrt{1 - |p_n|^2} < k_{\varphi(p_n)}, h > \\[2ex]
&= \quad \lim_n \sqrt{1 - |p_n|^2}\, \overline{h(\varphi(p_n))} \\[2ex]
&= \quad 0,
\end{aligned}
$$

where the third line follows from the Lemma at the end of the last section, and the last one from the fact that h is bounded on U. Thus g is orthogonal to every polynomial. Since the polynomials form a dense subset of H^2, it follows that g is the zero function.   □

With this result, the proof of the Univalent Compactness Theorem is complete.

3.6   Compactness and Contact

The results of Chapter 2 suggested a strong connection, at least for univalently induced composition operators, between compactness and the "degree of contact" that the image of the inducing map has with the unit circle.

It was shown, for example, that the operator is compact if this image is confined to an inscribed polygon, and non-compact if the image contains a disc tangent to the circle.

The Univalent Compactness Theorem allows us to considerably refine these results. In this section we show that a univalently induced composition operator will fail to be compact whenever, for some $\alpha > 1$, its image approaches the unit circle "faster than $y = x^\alpha$ approaches the real axis." In the other direction, we give an example that shows that the corners in the Polygonal Compactness Theorem (§2.3) can be rounded off "just a little" without loss of compactness.

Contact with the boundary. The first order of business is to decide how to measure the order of contact made by a region in U with the unit circle. For simplicity we consider only contact at the point $+1$; all the arguments work with obvious modifications for any other point of ∂U. Let $\gamma : [0, \pi] \to [0, 1)$ be a continuous function with $\gamma(0) = 0$, but $\gamma(\theta) > 0$ otherwise. We use γ to define a Jordan curve Γ in U by means of the polar equation

$$1 - r = \gamma(|\theta|) \qquad (|\theta| \le \pi).$$

Thus Γ is symmetric about the real axis, and lies in U except for a single point of intersection with the unit circle at $+1$. For a positive number α, let us agree to call Γ an α-curve at $+1$ if $\theta^{-\alpha}\gamma(\theta)$ has a (finite) non-zero limit as $\theta \to 0+$. For example, a triangle that is symmetric about the real axis and lies in U except for a vertex at $+1$ is a 1-curve, while a circle properly contained in $\overline{U}$, and tangent to ∂U at $+1$ is a 2-curve. Finally, we say Γ approaches the unit circle *smoothly* at $+1$ if $\theta^{-1}\gamma(\theta) \to 0$ as $\theta \to 0+$. Thus, every α-curve for $\alpha > 1$ approaches ∂U smoothly.

We say a region $\Omega \subset U$ has *contact* α with the unit circle at $+1$ if it contains an α-curve at $+1$ (we could be more precise and say Ω has contact *at least* α at $+1$). If Ω merely contains a curve that approaches the boundary smoothly at $+1$ then we say the region has *smooth contact* with ∂U at that point.

We will find it useful to express these definitions in terms of distances, both in the unit disc and the right half-plane.

Lemma 1. *Suppose* $\alpha \ge 1$. *Then* Γ *is an* α-curve *if and only if*

$$\lim \frac{1 - |z|}{|1 - z|^\alpha} \qquad (z \to 1, \ z \in \Gamma)$$

exists (finitely) and is non-zero.

Thus Γ is an α-curve if and only if for each of its points, the distance to the boundary is comparable to the α-th power of the distance to $+1$.

Proof. For $z = re^{i\theta} \in \Gamma$ we calculate

$$|1 - z|^2 = (1 - r)^2 + r \left(2 \sin \frac{\theta}{2} \right)^2,$$

hence for $|\theta| \to 0$,

$$\left(\frac{|1 - z|^\alpha}{1 - |z|} \right)^{2/\alpha} = \frac{\gamma(|\theta|)^2 + (1 + o(1))\theta^2}{\gamma(|\theta|)^{2/\alpha}}$$

$$= \gamma(|\theta|)^{2(1 - \frac{1}{\alpha})} + (1 + o(1)) \left(\frac{|\theta|}{\gamma(|\theta|)^{1/\alpha}} \right)^2.$$

The first summand of the last line is $\equiv 1$ if $\alpha = 1$, while if $\alpha > 1$ then it converges to 0 as $|\theta| \to 0$. This establishes our assertion.   □

We will be constructing univalent self-maps of the unit disc by working instead in the right half-plane Π, and then returning to the disc through the change of variable $w = \tau(z) = \frac{1+z}{1-z}$. Thus we need to know how the concept of "α-curve" fares under this change of scene. To make sense out of what is going to happen, it helps to keep in mind that τ transforms line segments through $+1$ into other line segments, but it also transforms circles tangent to ∂U at $+1$ into (vertical) lines.

Lemma 2. *Suppose γ and Γ are as above. Let $\tilde{\Gamma}$ be the image of γ under the map τ. Then Γ is an α-curve if and only if*

$$\lim \frac{\operatorname{Re} w}{|w|^{2-\alpha}} \qquad (w \to \infty, \ w \in \tilde{\Gamma})$$

exists and is non-zero.

Proof. The change of variable can be rewritten $z = \frac{w-1}{w+1}$, from which follows two important distance formulas:

$$1 - z = \frac{2}{w + 1} \qquad \text{and} \qquad 1 - |z|^2 = \left| \frac{2}{w + 1} \right|^2 \operatorname{Re} w. \qquad (8)$$

These show that as $z \to 1$, (equivalently: as $w \to \infty$),

$$\frac{1 - |z|}{|1 - z|^\alpha} = \frac{1}{1 + |z|} \frac{1 - |z|^2}{|1 - z|^\alpha}$$

$$= \left(\frac{1}{2} + o(1) \right) \left| \frac{2}{w + 1} \right|^{2-\alpha} \operatorname{Re} w$$

$$= 2^{1-\alpha}(1 + o(1)) \frac{\operatorname{Re} w}{|w|^{2-\alpha}},$$

which yields the desired result.   □

A class of examples. We can now write down for each $1 < \alpha < 2$ examples of univalent maps ψ for which $\psi(U)$ is a Jordan domain whose boundary is, near the point $+1$, an α curve, and for which C_ψ is not compact on H^2. We introduce two additional parameters $a, b > 0$ for later use (inviting the reader to set them both equal to 1 in the proof below), and define

$$\Psi(w) = \Psi_{\alpha,a,b}(w) = a + w + bw^{2-\alpha}$$

where the principal branch of the argument is used to define the fractional power on the right. Clearly Ψ maps the right half-plane into itself. Let $\psi = \psi_{\alpha,a,b}$ be the corresponding holomorphic self-map of U .

Proposition. *For each $1 < \alpha < 2$ and $a, b > 0$ the map ψ has these properties:*

(a) *ψ is univalent on $\overline{U}$, and $\psi(\overline{U}) \subset U \cup \{1\}$.*

(b) *$\psi(\partial U)$ is an α-curve at $+1$.*

(c) *C_ψ is not compact on H^2.*

Proof. (a) We work in the right half-plane. Univalence follows from the fact that the derivative of Ψ has positive real part. More precisely note that

$$\Psi'(w) = 1 + \frac{(2-\alpha)b}{w^{\alpha-1}}$$

has positive real part in $\overline{\Pi}\backslash\{0\}$, hence if w_1 and w_2 are distinct points of that set, and L is the line segment joining the points,

$$\Psi(w_2) - \Psi(w_1) = \int_L \Psi'(\zeta)d\zeta = (w_2 - w_1)\int_0^1 \Psi'(tw_2 + (1-t)w_1)dt.$$

Thus

$$\mathrm{Re}\left(\frac{\Psi(w_2) - \Psi(w_1)}{w_2 - w_1}\right) = \int_0^1 \mathrm{Re}\,\Psi'(tw_2 + (1-t)w_1)dt.$$

Since the integrand in the last integral is strictly positive, so is the integral, hence $\Psi(w_1) \neq \Psi(w_2)$. The argument works as well for $w_1 = 0$ since the singularity in the derivative is integrable, and clearly Ψ extends continuously to $\widehat{C}$, with $\Psi(\infty) = \infty$.

(b) By the definition of Ψ, we have for each real y,

$$\mathrm{Re}\,\Psi(iy) = a + bc(\alpha)|y|^{2-\alpha}$$

where $c(\alpha) = \cos(\frac{\pi}{2}(2-\alpha))$, which, because $0 < 2 - \alpha < 1$, is a positive number. Now the definition of Ψ shows that $|\Psi(iy)| = (1 + o(1))|y|$ as $|y| \to \infty$, hence

$$\lim_{|y|\to\infty} \frac{\mathrm{Re}\,\Psi(iy)}{|\Psi(iy)|^{2-\alpha}} = bc(\alpha) \tag{9}$$

This result, along with Lemma 2 above shows that the boundary of $\psi(U)$ is an α curve at $+1$.

(c) To prove that C_ψ is not compact on H^2 it suffices, by the Univalent Compactness Theorem of the last chapter, to show that as $x \to 1$ (in the unit interval) we have

$$\liminf_{x \to 1-} \frac{1 - \psi(x)}{1 - x} < \infty,$$

This is easy: for $0 < x < 1$ set $u = (1+x)/(1-x)$, the corresponding point of the positive real axis, and use the first of formulas (8) to calculate:

$$\frac{1 - \psi(x)}{1 - x} = \frac{2}{1 + \Psi(u)} \cdot \frac{1 + u}{2}$$

$$= \frac{1 + u}{a + 1 + u + bu^{2-\alpha}}.$$

Since $1 < \alpha < 2$ the last expression converges to 1 as $u \to \infty$, that is, as $x \to 1$.   □

With the help of the Comparison Theorem, we can use this class of examples to considerably refine the "tangent disc" result of §2.5.

Improved Non-compactness Theorem. *If φ is univalent and $\varphi(U)$ has contact $\alpha > 1$ with the unit circle at some point, then C_φ is not compact on H^2.*

Proof. Without loss of generality we may assume that the point of contact is $+1$. Thus we are assuming that the image of φ contains an α- curve at $+1$. By the Comparison Principle of §2.5 it suffices to prove that the Riemann map of the unit disc onto the region bounded by this α curve induces a non-compact composition operator. So without loss of generality we may assume that φ is this Riemann map. By Lemmas 1 and 2, and (9) above, we can choose the parameter b small enough so that the part of the boundary of $\psi_{\alpha,0,b}(U)$ that lies in some neighborhood V of $+1$ is contained in $\varphi(U)$.

We claim that a sufficiently large choice of the "translation" parameter a forces $\psi = \psi_{\alpha,a,b}$ to map U into V, and therefore completely into $\varphi(U)$, at which point the non-compactness of C_ψ will guarantee, via the Comparison Principle, that C_φ is not compact.

The argument is best visualized in the right half-plane, where the boundary of $\varphi(U)$ becomes a simple curve in Π that is symmetric about the real axis and heads out to ∞, and the neighborhood V becomes a neighborhood (which we still denote by V) of ∞, i.e. the exterior of some half-disc with center at the origin. Let $\Omega = \psi_{\alpha,0,b}(U)$, and write Φ for the counterpart of φ acting on Π. We have previously chosen the dilation parameter b so that $\Omega \cap V \subset \Phi(\Pi)$. Now it is easy to check that Ω is taken into itself by horizontal translation. Indeed, Ω is symmetric about the real axis and its upper boundary has the form $y = A + Bx^\gamma$, where $A, B > 0$ and $\gamma > 1$. The translation property follows from this symmetry and the fact that the upper boundary curve is the graph of a monotone increasing function.

Now choose $a > 0$ so that $a + \Omega \subset V$. Then by the work of the last paragraph,

$$\Psi_{\alpha,a,b}(\Pi) = a + \Omega \subset \Phi(\Pi),$$

and the proof is complete. $\square$

We close this chapter by using the Univalent Compactness Theorem to construct an example of a compact composition operator whose inducing map takes the disc onto a domain that touches the boundary smoothly.

Example of "smooth compactness." *There exist univalent self-maps φ of U such that:*

(a) $\varphi(\overline{U}) \subset U \cup \{+1\}$,

(b) $\varphi(U)$ *contacts ∂U smoothly at $+1$, and*

(c) C_φ *is compact on H^2.*

Proof. Instead of the unit disc, we work in the half-disc

$$\Delta = \{w \in \Pi : |w| < \frac{1}{2e}\},$$

with the holomorphic function

$$f(w) = -cw \log w \qquad (w \in \Delta)$$

where $c > 0$, and the principal branch of the logarithm is employed on the right. One easily checks that f maps Δ into a bounded subset of the right half-plane, so the constant c can be chosen so that $f(\Delta) \subset \Delta$. Moreover, f' has positive real part on $\overline{\Delta}\backslash\{0\}$, so as in the proof of the Proposition, f is univalent on $\overline{\Delta}$.

Now let τ be a univalent map taking Δ onto U, with $\tau(0) = +1$. This map extends to a homeomorphism of the corresponding closed regions— as can be seen by either writing it down as a composition of elementary mappings, or quoting Carathéodory's Extension Theorem [Rdn '87, §14.19–20]. Thus, the Reflection Principle insures that τ extends analytically to a mapping that takes the entire disc $|w| < 1/2e$ univalently onto a simply connected domain containing both U and the point $+1$.

Let φ be the univalent self-map of U that corresponds, via τ, to f on Δ. Since τ is analytic with non-vanishing derivative in a full neighborhood of the origin, distance estimates in Δ transfer over to corresponding distance estimates in U (we leave the precise formulation and proof of this statement to the reader). In particular, if $|y| < 1/2e$ and $e^{i\theta} = \tau(iy)$, then

$$\frac{1 - |\varphi(e^{i\theta})|}{|1 - \varphi(e^{i\theta})|} = \frac{\text{dist}\,(\varphi(e^{i\theta}), \partial U)}{\text{dist}\,(\varphi(e^{i\theta}), +1)}$$

$$\leq \text{const.}\,\frac{\text{dist}\,(f(iy), \partial\Pi)}{\text{dist}\,(f(iy), 0)}$$

$$= \text{const.}\,\frac{\text{Re}\,f(iy)}{|f(iy)|}$$

$$= \text{const.}\,\frac{\pi|y|/2}{\{(\pi|y|/2)^2 + (|y|\log|y|)^2\}^{1/2}}$$

$$\leq \frac{\text{const.}}{-\log|y|}$$

$$\to 0$$

as $|y| \to 0$, i.e., as $|\theta| \to 0$. Thus the image of the unit disc under φ approaches the boundary smoothly at $+1$.

The compactness of C_φ on H^2, will follow from the Univalent Compactness Theorem once we show that

$$\frac{1 - |\varphi(z)|}{1 - |z|} \to \infty \tag{10}$$

as z tends to any point of ∂U. We need only check this limit at the point $+1$, since the closure of $\varphi(U)$ approaches the unit circle nowhere else. For $z \in U$ let $z = \tau(w)$, and estimate as above,

$$\frac{1 - |\varphi(z)|}{1 - |z|} \geq \text{const.}\,\frac{\text{dist}\,(f(w), \partial\Pi)}{\text{dist}\,(w, \partial\Pi)}$$

$$= \text{const.}\,\frac{\text{Re}\,f(w)}{\text{Re}\,w}$$

$$= \text{const.}\,\frac{(\text{Re}\,w)(-\log(|w|)) + (\text{Im}\,w)\arg w}{\text{Re}\,w}$$

$$\geq \text{const.}\,\log(1/|w|),$$

where the last inequality follows from our use of the principal branch of the argument, which insures that both argument and imaginary part always have the same sign. Thus (10) holds as $z \to +1$ (equivalently, as $w \to 0$), as desired.   □

The idea behind this proof can be modified to produce Hilbert-Schmidt composition operators induced by univalent mappings that take the unit disc onto subdomains that contact the boundary smoothly (see Exercise 12 below). Used in conjunction with the Comparison Principle of §2.3 these examples produce "rounded corners" versions of the Polygonal Compactness Theorem. In this matter we opt for self-restraint and leave the formulation and proof of such theorems to the interested reader.

3.7 Exercises

1. *The Littlewood-Paley Identity.* Show that for each $f \in H(U)$,

$$\|f\|^2 = |f(0)|^2 + 2 \int_U |f'(z)|^2 \log \frac{1}{|z|} dA(z).$$

2. Show that for φ a *univalent* self-map of U, the Hilbert-Schmidt condition (3) of Chapter 2 holds whenever:

$$\int_{\varphi(U)} \frac{dA(w)}{(1 - |w|^2)^2} < \infty.$$

(This integral is called the "hyperbolic area" of $\varphi(U)$; see Chapter 9, Exercise 10 for further details.)

3. Use the above result to find examples of univalently induced compact composition operators C_φ for which the closure of $\varphi(U)$ touches the boundary at infinitely many points. Can this happen at uncountably many points? Can the whole unit circle lie in this closure?

4. Use the Univalent Compactness Theorem to show that:

   (a) The "lens maps" of §2.3 induce compact composition operators.

   (b) The map $(1+z)/2$ induces a non-compact composition operator.

5. *Generalization of Exercise 4(b).* Show that: If for some $\omega \in \partial U$

$$\lim_{r \to 1-} \varphi(r\omega) = \eta \in \partial U \quad \text{and} \quad \lim_{r \to 1-} \varphi'(r\omega) \text{ exists,}$$

   then C_φ is not compact on H^2.

6. *Finitely-valent Compactness Theorem.* Show that the conclusion of the Univalent Compactness Theorem still holds if φ is merely assumed to be finitely valent.

7. Construct a "ribbon mapping" that takes U onto $U\backslash\{0\}$ at most two to one, but induces a compact operator on H^2. *Suggestion:* Use the exponential map to get a ribbon that covers everything but the origin. Use Exercise 6 to conclude that the induced composition operator is compact.

8. Modify the example above to create a holomorphic function that takes U *onto* itself, but induces a compact composition operator (for details, see [MlS '86]).

9. *The compactness problem for the Bergman space.* Write down an appropriate representation of the norm in the Bergman space A^2 (see Exercise 4 of Chapter 1), and use it to prove that condition (2) of this chapter is also sufficient for compactness of univalently induced composition operators acting on the Bergman space. (See the next problem, and the *Notes* below for more on this topic.)

10. Determine the reproducing kernels for the Bergman space, and use them to show that (2) is also necessary for an composition operator to be compact on A^2 (univalence is not needed here).

11. Suppose φ is a univalent self-map of U that has radial limits of modulus 1 on a subset of the unit circle having positive measure. Show that for some $\omega \in \partial U$,

$$1 - |\varphi(r\omega)| = O(1-r) \quad \text{as} \quad r \to 1 - .$$

12. *More "smoothly induced" compact operators.* Regarding the example of "Smooth Compactness" given in the last section, show that for any $\alpha > 0$ the holomorphic function $f(w) = w(-\log w)^\alpha$ can be used to construct a univalent map that takes U onto a Jordan subdomain that approaches the unit circle at exactly one point, does so smoothly, and induces a compact composition operator. Show further that this operator is Hilbert-Schmidt (§2.3) if and only if $\alpha > 2$.

3.8   Notes

The Univalent Compactness Theorem was proved in [MlS '86]. In that paper condition (2) was shown to characterize the compact operators on the Bergman space A^2 introduced in Exercise 4 of §1.4 *without any extra assumption of univalence* (see Exercises 10 through 15 of Chapter 10 for a proof). In the setting of H^2, condition (2) remains sufficient for compactness under much weaker assumptions than univalence. For example, it is enough to assume C_φ is a bounded operator on the *Dirichlet space* introduced here in Exercise 9 of §1.4 ([MlS '86, Theorem 3.10]). According to that exercise, every univalent self-map of U has this property, so the above result implies our Univalent Compactness Theorem.

However, in Chapter 10 we will show that condition (2) is in general *not* sufficient for compactness on H^2, and we will solve the compactness problem for arbitrary holomorphic self-maps of U. For the reader who wishes to look at this right now, the only additional prerequisite is Jensen's Formula for the distribution of zeros of a holomorphic function, which is developed here in §7.3.

The work of §3.6 was adapted from [STa '73, §4], as was Exercise 12. To see how the results of this chapter fare in the setting of the unit ball of $\mathbf{C}^n$, see [Mlr '85].

4

The Angular Derivative

At the end of the last chapter we managed to coax a fair amount of geometric intuition out of the condition

$$\lim_{|z|\to 1-} \frac{1 - |\varphi(z)|}{1 - |z|} = \infty, \tag{1}$$

which characterizes compactness for univalently induced composition operators. Because this condition involves the limit of a difference quotient, one might suspect that its real meaning is wrapped up in the boundary behavior of the *derivative* of φ. This is exactly what happens: we will see shortly that condition (1) is the hypothesis of the classical Julia-Carathéodory Theorem, which characterizes the existence of the "angular derivative" of φ at points of ∂U, and provides a compelling geometric interpretation of (1) in terms of "conformality at the boundary." After discussing its connection with the compactness problem, we present a proof of the Julia--Carathéodory Theorem that emphasizes the role of hyperbolic geometry. The following terminology describes the limiting behavior involved in this circle of ideas.

Definition. (a) A *sector* (in U) at a point $\omega \in \partial U$ is the region between two straight lines in U that meet at ω and are symmetric about the radius to ω.

(b) If f is a function defined on U and $\omega \in \partial U$, then

$$\angle \lim_{z \to \omega} f(z) = L$$

means that $f(z) \to L$ as $z \to \omega$ through any sector at ω. When this happens, we say L is the *non-tangential* (or *angular*) limit of f at ω.

4.1 The Definition

We say a holomorphic self-map φ of U has an *angular derivative* at $\omega \in \partial U$ if for some $\eta \in \partial U$,

$$\angle \lim_{z \to \omega} \frac{\eta - \varphi(z)}{\omega - z}$$

exists (finitely). We call the limit the *angular derivative* of φ at ω, and denote it by $\varphi'(\omega)$.

Warning. The existence of the angular derivative implies that φ has angular limit η at ω. We are requiring that η be a point of the unit circle, so regardless of how smooth φ may be at the boundary, our definition demands:

> φ cannot have an angular derivative at any boundary point at
> which it fails to have an angular limit of modulus one.

Although this extra requirement on the angular limit might be overly restrictive for some purposes, for *ours* it is just right. The work so far shows that for the compactness problem, the important phenomena occur as $\varphi(z)$ approaches the boundary. For example, according to our definition, the function $\varphi(z) = z/2$ (which induces a compact composition operator) has an angular derivative *nowhere* on ∂U. In the same vein, observe that the lens maps introduced in Chapter 2, which also induce compact composition operators, do not have angular derivatives at any boundary points, while the map $\varphi(z) = (1 + z)/2$, which has an angular derivative at the point $+1$ (and, according to our definition, nowhere else) induces a non-compact operator.

These examples raise the possibility that the results of the last chapter might be restated in terms of the angular derivative. This is exactly what is going to happen. The "necessary" part of the program follows immediately from the definitions, and as before, does not require univalence.

Proposition. *If φ has an angular derivative at a point $\omega \in \partial U$ then C_φ is not compact on H^2.*

Proof. Letting η denote the angular limit of φ at ω we have

$$\liminf_{|z| \to 1-} \frac{1 - |\varphi(z)|}{1 - |z|} \le \liminf_{r \to 1-} \frac{1 - |\varphi(r\omega)|}{1 - r} \le \lim_{r \to 1-} \left| \frac{\eta - \varphi(r\omega)}{\omega - r\omega} \right| = |\varphi'(\omega)|,$$

so the result follows from the necessary condition for compactness derived in §3.5. $\qquad\square$

4.2 The Julia-Carathéodory Theorem

In addition to raising the issue of compactness, the definition of the angular derivative suggests that φ has some kind of conformality at the boundary points where it exists, and it further raises the possibility that the derivative of φ might also have a non-tangential limit at ω. The main theorem of this chapter asserts that all of these concepts are intimately bound up with each other.

The Julia-Carathéodory Theorem. *Suppose φ is a holomorphic self-map of U, and $\omega \in \partial U$. Then the following three statements are equivalent:*

(JC 1) $\displaystyle \liminf_{z \to \omega} \frac{1 - |\varphi(z)|}{1 - |z|} = \delta < \infty,$

(JC 2) $\displaystyle \angle \lim_{z \to \omega} \frac{\eta - \varphi(z)}{\omega - z}$ *exists for some $\eta \in \partial U$,*

(JC 3) $\displaystyle \angle \lim_{z \to \omega} \varphi'(z)$ *exists, and $\displaystyle \angle \lim_{z \to \omega} \varphi(z) = \eta \in \partial U$.*

Moreover:

- *$\delta > 0$ in (JC 1),*

- *the boundary points η in (JC 2) and (JC 3) are the same, and*

- *the limit of the difference quotient in (JC 2) coincides with that of the derivative in (JC 3), with both equal to $\omega\bar{\eta}\delta$.*

Since condition (JC 1) is just the compactness criterion of the last chapter, this allows the results of that chapter to be restated in terms of the angular derivative.

Angular Derivative Criterion for Compactness. *Suppose φ is a holomorphic self-map of U.*

(a) *If C_φ is compact on H^2 then φ has an angular derivative at no point of ∂U.*

(b) *If φ is univalent and has no angular derivative at any point of ∂U, then C_φ is compact on H^2.*

To appreciate the purely function-theoretic power of the Julia-Carathé-odory Theorem, observe how, almost as an afterthought, it asserts that if, on a sequence $\{z_n\}$ of points in U that converges to a boundary point ω, the images $\varphi(z_n)$ tend to the boundary rapidly enough, then regardless of how sparse or tangential $\{z_n\}$ may be, the function φ must have a radial limit at ω. So even ignoring what it says about derivatives, the Julia-Ca-rathéodory Theorem already yields a non-trivial result about boundary behavior.

Remark. Condition (JC 2) implies that φ is "non-tangentially conformal" at ω. To understand this conformality, we may without loss of generality take $\omega = \eta$. Then (JC 1) and the fact that $\delta > 0$, make it possible to recycle the proof from elementary complex analysis that "holomorphic plus non-vanishing derivative implies conformal." The argument (left to the reader) yields this:

> If a smooth curve in U ends at a point $\omega \in \partial U$, at which it makes an angle $\alpha < \pi/2$ with the radius to that point, then the same is true of the image curve at the boundary point η.

In particular, the image of the radius itself meets the unit circle perpendicularly, and if two non-tangential curves intersect at ω at some angle, then their images intersect at the same angle.

First applications. As an example of what the Julia-Carathéodory Theorem contributes to the compactness problem, note how it clarifies the intuition behind the Polygonal Compactness Theorem: if φ takes U into a polygon inscribed in the unit circle, then at no vertex preimage does φ have the conformality demanded by the Angular Derivative Criterion. Because of our requirement that the angular derivative can only exist at points whose (radial) preimage is on the unit circle, it exists at none of the other points either, so C_φ is compact.

For another example, recall the assertion of the "First Compactness Theorem" of §2.2:

> C_φ is compact whenever $\|\varphi\|_\infty < 1$.

The Angular Derivative Criterion effortlessly provides a significant generalization.

Corollary. If φ is univalent and has no radial limit of modulus 1, then C_φ is compact.

Suppose, for example, that φ maps the unit disc onto a subdomain in the form of a ribbon that swirls out to the boundary without overlapping itself, and circles the origin infinitely often. Then $\|\varphi\|_\infty = 1$, but clearly φ has no radial limit at any point of the unit circle (the image of each radius captures the entire circle in its closure). With a little more care this example can be modified to provide a mapping that takes the unit disc *onto* itself, covers every point at most twice, and yet induces a compact composition operator on H^2 (cf. Problems 7 and 8 of Chapter 3). Thus for the Comparison Principle of §2.5, the hypothesis of univalence is absolutely essential.

Before beginning to prove the Julia-Carathéodory Theorem, we need to isolate the main issue. The "upward implications" of the theorem are routine. For example, if both function and derivative converge as in (JC 3), then you integrate the derivative to get (JC 2), obtaining in the process the same limit you had for the derivative. The implication (JC 2) $\Rightarrow$(JC 1) is immediate if you let z tend to ω along the radius to ω, take absolute values, and use the "reverse triangle inequality." This also shows that the quantity δ on the right-hand side of (JC 1) is dominated by the magnitude of the limit in (JC 2).

The implication (JC 2) $\Rightarrow$(JC 3) requires a little more care, but it does not pose a real problem, and will be dispatched later. Thus:

> The heart of the Julia-Carathéodory Theorem is the implication (JC 1) $\Rightarrow$(JC 2).

The proof of this implication will occupy most of the rest of the chapter. Central to the argument is a remarkable generalization of the Schwarz Lemma, in which the role of the fixed point at the origin is assumed by a pair of points *on the unit circle*. We begin with an intermediate result: a simple but powerful generalization of the Schwarz Lemma, in which the origin is replaced by an arbitrary pair of points in U.

4.3   The Invariant Schwarz Lemma

The subject of this section, also called the Schwarz-Pick Lemma, is what results when you subject the Schwarz Lemma to a conformal change of variable. In order to state the result efficiently, we need a conformally invariant way of measuring distance in the unit disc.

To get started, recall the special conformal automorphisms α_p introduced in §0.4,

$$\alpha_p(z) = \frac{p - z}{1 - \bar{p}z}.$$

Definition. The *pseudo-hyperbolic distance* between points p and q of U is:

$$d(p, q) = |\alpha_p(q)| = \left| \frac{p - q}{1 - \bar{p}q} \right|. \tag{2}$$

The pseudo-hyperbolic distance is actually a *metric* on U (see Exercise 2 below) that induces the usual Euclidean topology; however our work requires only the following easily verified observations:

(a) For each pair of points $p, q \in U$ we have $d(p, q) = d(q, p)$, and $d(p, q) <$
    1.

(b) $d(p, q) = 0 \iff p = q$.

(c) d is continuous when viewed as a real-valued function on $U \times U$.

(d) For each compact subset K of U,

$$\lim_{|q| \to 1-} \inf_{p \in K} d(p, q) = 1.$$

Property (d) asserts that the pseudo-hyperbolic distance from a point to
a fixed *compact set* tends to 1 as the point tends to the boundary. This
is obvious if K is a single point, and not difficult to prove in general. For
example, it is rendered perfectly transparent by the formula

$$1 - d(p, q)^2 = \frac{(1 - |p|^2)(1 - |q|^2)}{|1 - \bar{p}q|^2}, \tag{3}$$

which is itself the result of a straightforward calculation. This formula will
prove very useful in our further study of the pseudo-hyperbolic distance.

With the pseudo-hyperbolic distance in our corner, we can state a simple
but far-reaching generalization of the Schwarz Lemma.

The Invariant Schwarz Lemma. *If φ is a holomorphic self-map of U,
then for every pair of points $p, q \in U$ we have*

$$d(\varphi(p), \varphi(q)) \leq d(p, q).$$

*Moreover there is equality here for some pair of points if and only if there
is equality for all pairs, and this happens if and only if φ is a conformal
automorphism of U.*

Proof. Note that if $p = \varphi(p) = 0$ then we are talking about the original
statement of the Schwarz Lemma. Otherwise, let $b = \varphi(p)$ and consider
the map $\alpha_b \circ \varphi \circ \alpha_p$, which takes the disc into itself, and fixes the origin.
Upon applying the Schwarz Lemma to this map, evaluating at the point
$z = \alpha_p(q)$, and noting that the automorphism α_p is its own inverse, we
obtain the inequality

$$|\alpha_b \circ \varphi(q)| \leq |\alpha_p(q)|,$$

which is precisely what we want. The case of equality follows from the
corresponding part of the original Schwarz Lemma. $\square$

   This form of the Schwarz Lemma asserts that holomorphic self-maps of U that are not automorphisms strictly decrease all pseudo-hyperbolic distances. To make a geometric statement out of this we have to examine the balls associated with the pseudo-hyperbolic distance. For $p \in U$ and $0 < r < 1$ the *r-ball centered at p* is

$$\Delta(p, r) = \left\{ z : \left| \frac{z - p}{1 - \bar{p}z} \right| < r \right\} = \alpha_p(rU),$$

(where, once again, we appeal to the fact that α_p is self-inverse).

   We call $\Delta(p, r)$ the *pseudo-hyperbolic disc of (pseudo-) center p and (pseudo-) radius r*. Since $\Delta(p, r)$ is the image of the disc rU under a conformal automorphism of U, it is an ordinary open disc. Its Euclidean dimensions (which we do not need here) can be determined writing down its definition and completing the square, or by using the conformality of disc automorphisms (see also Exercise 1 below). In terms of these discs, the statement of the Invariant Schwarz Lemma becomes

$$\varphi(\Delta(p, r)) \subset \Delta(\varphi(p), r).$$

If $p = 0$ the assertion is that $\varphi(rU) \subset rU$, which is of course just the geometric interpretation of the original Schwarz Lemma.

4.4   A Boundary Schwarz Lemma

In this section we push the Invariant Schwarz Lemma "out to the boundary." The idea is to examine pseudo-hyperbolic discs whose centers tend to a point ω of the unit circle, and whose radii tend to one, and find the condition on centers and radii that guarantees the convergence of such a family of discs to a disc tangent to the unit circle at ω.

   Equation (3) for the pseudo-hyperbolic distance, allows the definition of pseudo-hyperbolic disc to be rewritten as:

$$\Delta(p, r) = \{ z : |1 - \bar{z}p|^2 < \frac{1 - |p|^2}{1 - r^2}(1 - |z|^2) \}. \tag{4}$$

Now equation (4) really wants to tell us something about the limiting behavior of discs. For if p tends to a point ω on ∂U, and r tends to 1 in such a way that

$$\frac{1 - |p|}{1 - r} \to \lambda \in (0, \infty),$$

then the expression on the right-hand side of the inequality in (4) converges to $\lambda(1 - |z|^2)$, while the one on the left goes to $|1 - z\bar{\omega}|^2$. Therefore $\Delta(p, r)$ must be converging (somehow) to the set $H(\omega, \lambda)$ defined by

$$H(\omega, \lambda) \stackrel{\text{def}}{=} \{ z : |1 - z\bar{\omega}|^2 < \lambda(1 - |z|^2) \}. \tag{5}$$

Upon completing the square in the inequality on the right, we find that $H(\omega, \lambda)$ is the Euclidean disc centered at $\omega/(1 + \lambda)$, of radius $\lambda/(1 + \lambda)$. In particular, $H(\omega, \lambda)$ is tangent to ∂U at the point ω, it expands as λ increases, and as $\lambda \to \infty$ it fills up the whole unit disc. We call $H(\omega, \lambda)$ a *horodisc* at ω.

Figure 4.1 below shows the evolution of $\Delta(p, r)$ as p tends to the boundary, with p and r related as above. Of course our assertion that $\Delta(p, r) \to H(\omega, \lambda)$ needs to be interpreted precisely, and this is best done in terms of sequences. We state the result formally, and leave the proof to the reader.

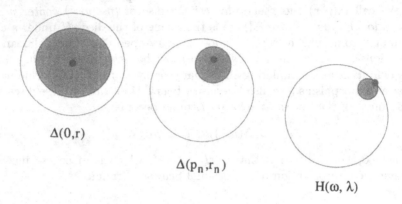

$\Delta(0,r)$

$\Delta(p_n,r_n)$

$H(\omega, \lambda)$

FIGURE 4.1. *Evolution of $\Delta(p, r)$ to a horodisc.*

The Disc Convergence Lemma. *Suppose $\omega \in \partial U$, and $\{p_n\}$ is a sequence of points in U that converges to ω. Suppose $0 < r_n \to 1$ in such a way that*

$$\lambda = \lim_n \frac{1 - |p_n|}{1 - r_n}.$$

Then

$$H(\omega, \lambda) \subset \liminf_n \Delta(p_n, r_n) \subset \limsup_n \Delta(p_n, r_n) \subset \overline{H(\omega, \lambda)}.$$

Here the lim sup of a sequence of sets is the collection of points that belong to infinitely many of the sets, and the corresponding lim inf is the collection of points that belong to all the sets from some index onward (the points that are "eventually in all the sets"). The Lemma says that if you treat the converging discs as if they behaved like points, your errors will be confined to boundary points.

Now we are in position to use the Disc Convergence Lemma to get a boundary version of the Invariant Schwarz Lemma.

Julia's Theorem. *Suppose φ is a non-constant holomorphic self-map of U, and that η and ω are points of ∂U. Suppose further that $\{p_n\}$ is a sequence of points in U that converges to ω in such a way that both $\varphi(p_n) \to \eta$ and*

$$\frac{1 - |\varphi(p_n)|}{1 - |p_n|} \to \delta < \infty. \tag{6}$$

Then:

(a) $\delta > 0$,

(b) $\varphi(H(\omega, \lambda)) \subset H(\eta, \lambda\delta)$ *for every* $\lambda > 0$, *and*

(c) $\angle \lim_{z \to \omega} \varphi(z) = \eta$.

Proof. (a) We first show that $\delta > 0$. Note that if φ were to fix the origin then the Schwarz Lemma would tell us straightaway that $\delta \geq 1$. The same idea works when we apply the Invariant Schwarz Lemma with $q = 0$, and it yields

$$d(\varphi(p), \varphi(0)) \leq d(p, 0) = |p|$$

for every $p \in U$. Upon rewriting this inequality using the identity (3), and doing a little algebra, we obtain

$$\frac{|1 - \overline{\varphi(p)}\varphi(0)|^2}{1 - |\varphi(0)|^2} \leq \frac{1 - |\varphi(p)|^2}{1 - |p|^2}.$$

Now the triangle inequality shows that

$$\frac{1 - |\varphi(0)|}{1 + |\varphi(0)|} \leq \frac{|1 - \overline{\varphi(p)}\varphi(0)|^2}{1 - |\varphi(0)|^2},$$

and upon putting last two inequalities together, setting $p = p_n$ and letting $n \to \infty$, we obtain

$$\frac{1 - |\varphi(0)|}{1 + |\varphi(0)|} \leq \frac{1 - |\varphi(p_n)|^2}{1 - |p_n|^2} \to \delta,$$

hence $\delta > 0$, as desired.

(b) Fix $0 < \lambda < \infty$. We may assume that $1 - |p_n| < \lambda$ for all n (if $\lambda \geq 1$ this is automatic, while if $\lambda < 1$ it can be accomplished, because $|p_n| \to 1-$, by discarding finitely many terms). Then all of the numbers

$$r_n = 1 - \frac{1 - |p_n|}{\lambda}$$

belong to the open unit interval, $r_n \to 1$, and

$$\frac{1 - |p_n|}{1 - r_n} = \lambda$$

for all n. Thus the sequence of discs $\Delta(p_n, r_n)$ satisfies the hypothesis of the Disc Convergence Lemma.

Now our hypothesis (6) insures that

$$\frac{1 - |\varphi(p_n)|}{1 - r_n} \to \lambda\delta$$

so the discs $\Delta(\varphi(p_n), r_n)$ also satisfy the hypothesis of the Disc Convergence Lemma, this time with η in place of ω and $\lambda\delta$ in place of λ. That Lemma and the Invariant Schwarz Lemma yield

$$\varphi(H(\omega, \lambda)) \subset \limsup_n \varphi(\Delta(p_n, r_n))$$

$$\subset \limsup_n (\Delta(\varphi(p_n), r_n))$$

$$\subset \overline{H(\eta, \lambda\delta)}.$$

Being non-constant, φ is an open mapping, so the closure symbol can be removed from the last term above, resulting in just the statement we wanted to prove.

(c) Suppose S is a sector in U with vertex at ω. We want to show that $\varphi(z) \to \eta$ as $z \to \omega$ through S. Given $\varepsilon > 0$, choose λ so that $H(\eta, \delta\lambda)$ is contained in the ε-disc centered at η. A quick sketch shows that a positive number ρ can be chosen so that the intersection of S with the ρ-disc centered at ω lies in $H(\omega, \lambda)$. By part (b),

$$z \in S \quad \text{and} \quad |z - \omega| < \rho \Rightarrow |\varphi(z) - \eta| < \varepsilon,$$

which is the desired result. $\square$

In fact the non-tangential convergence here can be improved to a certain kind of *tangential* convergence; see Exercise 7 below. Note also that Julia's Theorem gives a useful conclusion about the "boundary image point" η in condition (JC 2).

Corollary. *If $\{p_n\}$ is a sequence in U that converges to $\omega \in \partial U$, and on which the quotients $(1 - |\varphi(p_n)|)/(1 - |p_n|)$ are bounded, then $\{\varphi(p_n)\}$ converges to some point $\eta \in \partial U$, and φ has angular limit η at ω.*

Proof. Since $|\varphi(p_n)| \to 1$, each subsequence of the image sequence $\{\varphi(p_n)\}$ has a further subsequence that converges to some point of ∂U that Julia's Theorem asserts is then the angular limit of φ. So there is just one such subsequential limit point, which must therefore be the limit of the image sequence. $\square$

4.5   Proof that (JC 1) $\Rightarrow$ (JC 2)

The hypothesis (JC 1) asserts that we can choose a sequence $\{z_n\}$ in U
such that

$$z_n \to \omega \quad \text{and} \quad \frac{1 - |\varphi(z_n)|}{1 - |z_n|} \to \delta.$$

The existence of the boundary image point $\eta = \varphi(\omega)$ that shows up in (JC
2) was established in the last paragraph. In particular $\angle \lim_{z \to \omega} \varphi(z) = \eta$,
and the conclusion of Julia's Theorem further asserts that

$$\varphi(H(\omega, \lambda)) \subset H(\eta, \delta\lambda) \quad \text{for every} \quad \lambda > 0. \tag{7}$$

After appropriate preliminary rotations, we may assume that $\omega = \eta = +1$. The implications of Julia's Theorem for our result become more transparent if we replace the unit disc by the right half-plane Π, letting the point
at ∞ play the role of 1. This is accomplished by the mapping $\tau : \Pi \to U$
defined by

$$z = \tau(w) = \frac{w - 1}{w + 1} \quad (w \in \Pi).$$

The half-plane version of φ is $\Phi = \tau^{-1} \circ \varphi \circ \tau$. As we observed in §3.6,
distances in the disc and the half-plane are related by the equations

$$1 - z = \frac{2}{1 + w}, \tag{8}$$

and

$$1 - |z|^2 = \frac{4\operatorname{Re} w}{|w + 1|^2}. \tag{9}$$

These equations, along with (5), show that the horodisc $H(1, \lambda)$ in U corresponds, via τ, to the halfplane

$$\Pi(\lambda) = \{\operatorname{Re} w > \frac{1}{\lambda}\},$$

so the disc-mapping result (7) above can be rewritten as a condition on
half-planes: $\Phi(\Pi(\lambda)) \subset \Pi(\lambda\delta)$. Expressing this analytically,

$$\operatorname{Re} \Phi(w) \geq \frac{1}{\delta} \operatorname{Re} w \quad (w \in \Pi). \tag{10}$$

By (8), the quotient on the left side of (JC 2) can be written

$$\frac{1 - \varphi(z)}{1 - z} = \frac{w + 1}{\Phi(w) + 1}. \tag{11}$$

But $\angle \lim_{z \to 1} \varphi(z) = 1$, so $\Phi(w) \to \infty$ whenever $w \to \infty$ through any fixed
sector in Π, symmetric about the real axis, with angular opening $< \pi$ (in

symbols: $\angle \lim_{w\to\infty} \Phi(w) = \infty$). Thus (11) reveals the half-plane version of (JC 2) to be:

$$\angle \lim_{w\to\infty} \frac{\Phi(w)}{w} = \frac{1}{\delta}. \tag{12}$$

So it all comes down to proving this:

> If Φ is a holomorphic self-map of Π that satisfies (10), then it also satisfies (12).

There is one last reduction to be made. Let

$$c = \inf_{w\in\Pi} \frac{\operatorname{Re}\Phi(w)}{\operatorname{Re}w}.$$

Thus $\operatorname{Re}\Phi(w) \geq c\operatorname{Re}w$ for all $w \in \Pi$, and c is the largest number with this property. In view of (10) we must have $c \geq 1/\delta$. We are going to focus on the holomorphic function γ defined on Π by

$$\gamma(w) = \Phi(w) - cw.$$

The definition of c insures that γ has non-negative real part on Π, and that

$$\inf_{w\in\Pi} \frac{\operatorname{Re}\gamma(w)}{\operatorname{Re}w} = 0. \tag{13}$$

We are going to show that

$$\angle \lim_{w\to\infty} \frac{\gamma(w)}{w} = 0. \tag{14}$$

Since $\Phi(w) = cw + \gamma(w)$ on Π, this will show that

$$\angle \lim_{w\to\infty} \frac{\Phi(w)}{w} = c, \tag{15}$$

which is the desired conclusion (12), except that the limit is c instead of $1/\delta$. But that's good enough, since (15) translates into (JC 2) with the limit having magnitude $1/c$. In view of (JC 1), this limit must have magnitude at least δ, hence $1/c \geq \delta$. But we just observed that the opposite inequality also holds, so $c = 1/\delta$. Therefore we need only prove (14).

Proof of (14). This argument is entirely geometric. Recall that we are assuming γ is a holomorphic self-map of Π that satisfies (13). Fix an angle $0 < \beta < \pi/2$, and define the sector

$$S_\beta = \{w \in \Pi : |\arg w| < \beta\}.$$

We are supposed to show that $\gamma(w)/w \to 0$ as w tends to ∞ through this sector. To this end fix $\varepsilon > 0$. The goal is to find $R = R(\varepsilon) > 0$ so that

$$w \in S_\beta \quad \text{and} \quad |w| > R \quad \Rightarrow \quad \left|\frac{\gamma(w)}{w}\right| < \varepsilon.$$

For this, fix (forever) any number α with $\beta < \alpha < \pi/2$. Our fundamental assumption (13) guarantees a point $w_0 \in \Pi$ such that

$$\operatorname{Re}\gamma(w_0) < \varepsilon \operatorname{Re} w_0. \tag{16}$$

We initially choose R so that

$$w \in S_\beta \quad \text{and} \quad |w| > R \quad \Rightarrow \quad w \in w_0 + S_\alpha,$$

as shown in Figure 4.2.

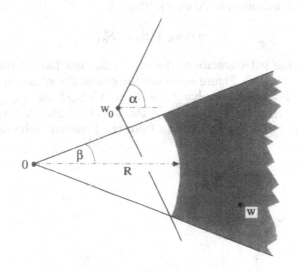

FIGURE 4.2. Sectors S_β and $w_0 + S_\alpha$, and radius R

With this in hand, the rest of the argument hinges on the Invariant Schwarz Lemma, which we transfer from the unit disc to the right half-plane by means of the map τ. To avoid notational proliferation, we now let $\Delta(p, r)$ denote the pseudo-hyperbolic disc in Π of (pseudo) radius r and (pseudo) center p. This means that if $q \in U$ is the image of $p \in \Pi$ under τ, then the disc we are calling $\Delta(p, r)$ in Π is precisely the one whose τ-image is $\Delta(q, r)$ in U. With this change of scenery, the Invariant Schwarz Lemma, when applied to the map γ, yields:

$$\gamma(\Delta(p, r)) \subset \Delta(\gamma(p), r).$$

The right half-plane offers the advantage that conformal changes of variable can be effected by affine automorphisms. For $\varphi \in \Pi$ let

$$A_p(w) = (\operatorname{Re}p)w + i\operatorname{Im}p \qquad (w \in \Pi).$$

Now A_p maps Π onto itself, and transfers the point 1 to p. The equality part of the Invariant Schwarz Lemma insures that any conformal automorphism of Π (or U) maps the pseudo-hyperbolic disc of radius r and center p onto the pseudo-hyperbolic disc of the same radius, and whose center is the image of the original center (where the words "radius" and "center" are always understood to carry the prefix "pseudo-"). Thus $A_p(\Delta(1,r)) = \Delta(p,r)$, that is:

$$\Delta(p,r) = (\operatorname{Re} p)\Delta(1,r) + i \operatorname{Im} p. \tag{17}$$

Now temporarily fix $w \in S_\beta$ with $w > R$, so that $w \in w_0 + S_\alpha$. Let r be the pseudo-hyperbolic distance from w_0 to w, so w is on the boundary of the pseudo-hyperbolic disc $\Delta(w_0, r)$. Thus

$$w \in w_0 + S_\alpha \subset S'_\alpha,$$

where S'_α is the left-translate of the sector S_α that has its vertex on the boundary of $\Delta(w_0, r)$. Figure 4.3 below illustrates the situation, and shows how crucial dimensions involving the discs $\Delta(w_0, r)$ and $\Delta(\gamma(w_0), r)$ are derived via (17) above from those of the standard disc $\Delta(1, r)$. The fact that $\gamma(w)$ lies in $\Delta(\gamma(w_0), r)$ follows from the Invariant Schwarz Lemma.

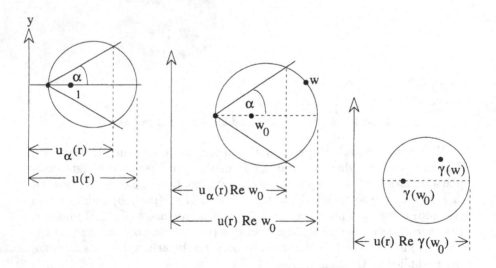

FIGURE 4.3. (l to r) The standard disc, $\Delta(w_0, r)$, and $\Delta(\gamma(w_0), r)$

We finish the proof of the Julia-Carathéodory Theorem by reading information off these diagrams. A little trigonometry applied to the picture of the standard disc shows that

$$u(r) \le (\sec^2 \alpha) u_\alpha(r), \tag{18}$$

which, used in conjunction with the information on other two diagrams, yields the following string of inequalities:

$$|\gamma(w) - \gamma(w_0)| \quad < \quad \operatorname{Re}\gamma(w_0)u(r)$$

$$< \quad \varepsilon(\operatorname{Re}w_0)u(r) \qquad\qquad \text{[by (16)]}$$

$$\leq \quad \varepsilon(\operatorname{Re}w_0)(\sec^2\alpha)u_\alpha(r) \qquad \text{[by (18)]}$$

$$\leq \quad (\varepsilon\sec^2\alpha)\operatorname{Re}w$$

Now replace $\operatorname{Re}w$ by $|w|$ in the last line and divide both sides of the resulting inequality by $|w|$. After a little further rearranging, aided by the triangle inequality, there results

$$\left|\frac{\gamma(w)}{w}\right| < \left|\frac{\gamma(w_0)}{w}\right| + \varepsilon\sec^2\alpha.$$

Upon letting w tend to ∞ through S_β we obtain

$$\limsup_{|w|\to\infty, w\in S_\beta} \left|\frac{\gamma(w)}{w}\right| \leq \varepsilon\sec^2\alpha.$$

Since ε was arbitrary, we have proved (14). This completes the implication (JC 1) ⇒(JC 2), the major part of the proof of the Julia-Carathéodory Theorem.

Remark. The proof above involved showing that (JC 1) implies that for every $a > 0$,

(JC $1\frac{1}{2}$) $\Phi(\Pi + a) \subset \Pi + \frac{a}{\delta}$

and that this geometric condition implies (JC 2). Thus we can add (JC $1\frac{1}{2}$) to our list of conditions equivalent to the existence of the angular derivative.

One final detail remains.

4.6 Proof that (JC 2) ⇒(JC 3)

As before, we may assume that $\omega = \eta = +1$. Thus we are assuming that $(1 - \varphi(z))/(1 - z)$ has limit δ as $z \to +1$ non-tangentially. We wish to show that $\varphi'(z)$ has non-tangential limit δ at $+1$. Let S be a sector in U with vertex at $+1$, and sides making angle α with the unit interval. We want to prove that

$$\lim \varphi'(z) = \delta \qquad (z \to 1, z \in S).$$

Let S' be a slightly larger sector with vertex at $+1$, and call its half-angular opening β, so $\alpha < \beta < \pi/2$. Let $\varepsilon > 0$ be given. Our hypothesis implies that

$$\varphi(z) - 1 = \delta(z - 1) + \psi(z) \tag{19}$$

where

$$\frac{\psi(z)}{1-z} \to 0$$

as z tends to 1 through S'.

The key is to represent the derivative of φ by the Cauchy formula, using the following contour of integration: For z in the smaller sector S, let $C(z)$ denote the circle with center z that is tangent to the boundary of S' at the side closest to z (see Figure (4.4) below). Let $r(z)$ be the radius of $C(z)$ (center and radius are "Euclidean" here).

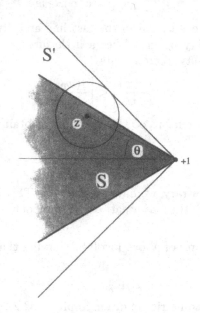

FIGURE 4.4. *Two sectors and a circle.*

Referring to this figure, we see that for each $z \in S$, if we write θ for the angle made by the segment $[z, 1]$ with the real axis, then

$$\frac{r(z)}{|1-z|} = \sin(\beta - \theta) \geq \sin(\beta - \alpha). \tag{20}$$

Successively using the Cauchy formula and (19) above, we obtain

$$\varphi'(z) = \frac{1}{2\pi i} \int_{C(z)} \frac{\varphi(\zeta) - 1}{(\zeta - z)^2} d\zeta$$

$$= \frac{1}{2\pi i} \int_{C(z)} \frac{\delta(\zeta - 1) + \psi(\zeta)}{(\zeta - z)^2} d\zeta$$

$$= \frac{\delta}{2\pi i} \int_{C(z)} \frac{d\zeta}{\zeta - z} + \frac{1}{2\pi i} \int_{C(z)} \frac{\delta(z - 1) + \psi(\zeta)}{(\zeta - z)^2} d\zeta$$

$$= \delta + \frac{1}{2\pi i} \int_{C(z)} \frac{\psi(\zeta)}{(\zeta - z)^2} d\zeta.$$

So everything depends on showing that the integral $I(z)$ in the last line tends to zero as z tends to 1 through S. Let $\varepsilon > 0$ be given. By hypothesis, if $\zeta \in S$ is sufficiently close to 1, then $|\psi(\zeta)| < \varepsilon |1 - \zeta|$. Thus if z is close enough to 1,

$$|I(z)| \leq \frac{\varepsilon}{2\pi} \int_{C(z)} \frac{|1 - \zeta|}{|\zeta - z|^2} |d\zeta|$$

$$\leq \frac{\varepsilon}{r(z)} \max_{\zeta \in C(z)} \{|1 - \zeta|\}$$

$$= \frac{\varepsilon}{r(z)} (r(z) + |1 - z|)$$

$$= \varepsilon \left(1 + \frac{|1 - z|}{r(z)} \right)$$

$$\leq \varepsilon(1 + \csc(\beta - \alpha)) \qquad \text{(from (20)).}$$

Since ε is an arbitrary positive number this shows that $I(z) \to 0$ as $z \to +1$ through S, and completes the proof that (JC 2) ⇒(JC 3). The proof of the Julia-Carathéodory Theorem is now complete.   □

So far our philosophy has been to use operator theory as a context for suggesting questions about function theory. Now we can turn the tables a bit (cf. Exercise 11 of §3.7).

Corollary *Suppose φ is a holomorphic self-map of U that has radial limits of modulus one on a set of positive measure in ∂U. Then φ has an angular derivative at some point of ∂U.*

Proof. According to the Proposition of §2.5, the operator C_φ is not compact on H^2, thus the Angular Derivative Criterion insures that it has an angular derivative somewhere on ∂U.   □

At the end of §2.5 we observed that C_φ will fail to be compact if the values of φ approach the unit circle either "too quickly" or "too often." The above Corollary asserts that if these values approaches the boundary "too often" for compactness, then at some boundary point they must also do so "too quickly."

4.7   Angular derivatives and contact

The results of §3.6 can be rephrased to relate the existence of the angular derivative of a univalent self-map of U to the way in which the map's image contacts the unit circle.

Proposition. *Suppose $\alpha > 1$ and Ω is a simply connected sub-domain of U that makes α-contact with ∂U at some point η. Suppose φ is a univalent function that maps U onto Ω. Then there is a point $\omega \in \partial U$ such that $\varphi(\omega) = \eta$ and φ has a finite angular derivative at ω.*

Proof. The image of $\psi = (\eta + \varphi)/2$ also makes α-contact with ∂U at η, and contacts the circle nowhere else. By the work of §3.6, ψ induces a non-compact composition operator, so the Univalent Compactness Criterion guarantees that ψ has an angular derivative at some point $\omega \in \partial U$. Now ψ has a radial limit $\psi(\omega) \in \partial U$. Because $\psi(U)$ lies in a horodisc at η, we must have $\psi(\omega) = \eta$. Thus also $\varphi(\omega) = \eta$, and a simple computation shows that φ also has an angular derivative at ω.   □

This proof can also be accomplished without flaunting the notion of compactness. The arguments used in §3.6 suffice, along with an "angular derivative" version of the chain rule (see Exercise 10 below).

The Proposition is, at its heart, a result about "α-curves." As such it is a very special case of a much deeper theorem whose proof we omit (see [Tsj '59, Theorems IX.9 and IX.10, pp.366–379] for details):

The Tsuji-Warschawski Theorem. *Suppose Ω is a Jordan subdomain of U whose boundary curve in a neighborhood of $+1$ has polar equation $1 - r = \gamma(|\theta|)$, where $\gamma : [0, \varepsilon] \to [0, 1]$ is a continuous, increasing function with $\gamma(0) = 0$. Let φ be a univalent map of U onto Ω, with $\varphi(1) = 1$. Then φ has an angular derivative at $+1$ if and only if*

$$\int_0^\varepsilon \frac{\gamma(\theta)}{\theta^2} d\theta < \infty.$$

Recent work of Burdzy and Rodin-Warchawski has refined this result into a complete solution of the problem of existence of the angular derivative; see the *Notes* at the end of this chapter for more details and references.

4.8   Exercises

1. *Euclidean dimensions of pseudo-hyperbolic discs.* (a) Suppose $p \in U$ and $0 < r < 1$. By considering the image of the disc $\{|a| < r\}$ under an appropriate member of $\mathrm{Aut}(U)$, show that the pseudo-hyperbolic

disc $\Delta(p, r)$ is the Euclidean disc with diameter on the line through the origin and p, and intersecting that line in the points

$$\frac{|p| - r}{1 - |p|r} \cdot \frac{p}{|p|} \quad \text{and} \quad \frac{|p| + r}{1 + |p|r} \cdot \frac{p}{|p|}$$

(note that it suffices to prove this for $0 < p < 1$).

(b) Use part (a) to give another proof that the Euclidean center and radius of $\Delta(p, r)$ are as given in §4.4 above.

2. *The triangle inequality for the pseudo-hyperbolic distance.*

(a) Show that if z, w, and p are points of U then

$$d(z, w) \leq \frac{d(z, p) + d(p, w)}{1 + d(z, p)d(p, w)}.$$

(b) Use the inequality of part (a) to derive the triangle inequality for the pseudo-hyperbolic distance.

Suggestion: For part (a), write $r = d(z, p)$ and use Exercise 1 above to conclude that

$$|z| \leq \frac{|p| + r}{1 + |p|r}.$$

Interpret this as the "$w = 0$" case of the desired inequality, and derive the general case by conformal invariance.

3. Discuss the case of equality in the triangle inequality derived in part (b) of the previous problem. In particular, given $z, w \in U$, is there a point $p \in U$ that is "half-way between z and w," in the sense that $d(z, p) = d(p, w) = d(z, w)/2$?

4. Exercise 2 above completes the proof that the pseudo-hyperbolic distance is a metric on U. Show that this metric is *not* complete.

5. *Julia-Carathéodory for automorphisms.* Show by direct computation that for each $\varphi \in \text{Aut}(U)$ and $w \in U$,

$$\lim_{z \to w} \frac{1 - |\varphi(z)|}{1 - |z|} = |\varphi'(w)|.$$

Suggestion: Compute instead the quantity $(1 - |\varphi(z)|^2)/(1 - |z|^2)$. Begin with φ a special automorphism, and use Exercise 2 of §0.5

6. Suppose φ is a holomorphic self-map of U that has a radial limit of modulus 1 at $\zeta \in \partial U$ and has bounded derivative on the radius from the origin to ζ. Show that φ has an angular derivative at ζ.

7. *Angular derivatives and* tangential *convergence.* By definition, if φ has an angular derivative at $\omega \in \partial U$ then it also has an angular limit there. More is true: Suppose Ω is a region in U bounded by a Jordan curve Γ that passes through ω, but otherwise lies in U. Suppose further that Γ is of class C^2, except at the point ω, and its curvature tends to ∞ as the curve approaches ω. Use the Julia-Carathéodory Theorem to show that $\varphi(z)$ has a limit as z approaches ω through Ω.

8. Suppose φ is a univalent self-mapping of the unit disc that extends continuously, but not univalently, to the boundary. Show that if points ω_1 and ω_2 of ∂U are mapped by φ to the same point on the unit circle, and φ has an angular derivative at ω_1, then φ *does not* have an angular derivative at ω_2. Give an example that shows univalence is needed here.

   Suggestion: Consider the images of triangles with vertices at the preimage points.

9. *Angular derivative of a product.* Suppose φ_1 and φ_2 are holomorphic self-maps of U, and $\varphi = \varphi_1 \varphi_2$. Show that if any two of these maps has an angular derivative at $\zeta \in \partial U$, then so does the third, in which case $|\varphi'(\zeta)| \geq |\varphi_j'(\zeta)|$ for $j = 1, 2$.

10. *The Chain Rule.* Suppose φ and ψ are holomorphic self-maps of U, and that φ has an angular derivative at ω, with $\varphi(\omega) = \eta$, and ψ has an angular derivative at η. Show that $\psi \circ \varphi$ has an angular derivative at ω, and find its value in terms of $\varphi'(\omega)$ and $\psi'(\omega)$.

11. Use the chain rule above to prove, without using operator theory, that the product of a compact, univalently induced composition operator with any composition operator is again compact.

12. Suppose $\{\varphi_n\}$ is a sequence of holomorphic self-maps of U that converges uniformly on compact subsets of U to φ. Suppose $\zeta \in U$, each φ_n has an angular derivative at ζ, and $\sup_n |\varphi'(\zeta)| < \infty$. Show that the limit function φ has an angular derivative at ζ.

   Suggestion: A review of the argument used to prove the Julia-Carathéodory Theorem shows that for φ any holomorphic self-map of U, another equivalent to $|\varphi'(\zeta)| = \delta$ is the global geometric condition that Φ, the right half-plane version of φ, map the half-plane $\{\operatorname{Re} w > a\}$ into the half-plane $\{\operatorname{Re} w > a/d\}$.

13. Show directly that the integral in the Tsuji-Warschawski Theorem diverges for the "smooth compactness" example of §3.6.

4.9 Notes

For Julia's Theorem, the foundation for the work of this chapter, see [Jla '20], and Carathéodory's contribution is in [Cth '29]. However the question "Who proved the Julia-Carathéodory Theorem?" is not a joke; others such as Wolff [Wf2 '26], Landau and Valiron [LVn '29], and the Nevanlinna brothers [Nvl '22] legitimately claimed significant parts of the action. The proof given here is based on Valiron's paper [Vln '31], which incorporates the work of [LVn '29], and provides some historical background. For more recent contributions, see Serrin [Srn '56] and Goldberg [Gbg '62]; the latter paper details concisely some of the "post-Julia" history of the result. In an interesting recent development, Sarason uses Hilbert space constructions due to de Branges and Rovnyak to give an operator-theoretic proof of the Julia-Carathéodory Theorem [Ssn '88].

Proofs of the Julia-Carathéodory theorem can be found in several books, for example [Cth '54, §295 – §303] and [Nvl '53, Ch. III, §5.3]. The result was generalized to the unit ball of $\mathbf{C}^n$ by Hervé [Hrv '63](see also [Rdn '80, §8.5]).

There is a long history associated with the problem of relating the existence of angular derivatives for univalent maps with the "degree of contact" their image boundaries have with the unit circle. The problem has only recently been solved, and the solution turns out to be a refinement of the Tsuji-Warchawski Theorem stated at the end of §4.7. To simplify matters we state the result in the upper half-plane (the setting in which it was proved), and leave it to the reader to formulate the equivalent version for the disc.

Suppose Ω is a simply connected domain in the upper half-plane, containing the origin in its boundary. Let $\mathcal{B}_\varepsilon$ be the collection of non-negative functions h on the real line for which:

(a) $h(0) = 0$.

(b) $|h(s) - h(t)| \leq |s - t|$ for all real s and t, and

(c) $\partial\Omega \cap \varepsilon U$ lies below the graph of h.

(Condition (b) says that each element of $\mathcal{B}_\varepsilon$ is a *Lipschitz function* on the real line, with Lipschitz constant 1; any other fixed Lipschitz constant would work as well.) Finally, let $h_\varepsilon(x) = \inf\{h(x) : h \in \mathcal{B}_\varepsilon\}$, so h_ε is a Lipschitz function whose graph is an "ε-majorant" for the boundary of Ω. In terms of these concepts, the solution of the angular derivative problem can be stated as follows.

Burdzy's Theorem. *Suppose Φ is a univalent mapping of the upper half-plane onto Ω, and that $\Phi(0) = 0$. Then Φ has an angular derivative at the* *origin if and only if for some $\varepsilon > 0$,*

$$\int_{-1}^{1} \frac{h_\varepsilon(x)}{x^2} dx < \infty.$$

This result is due to K. Burdzy [Bdz '86], who proved it using probabilistic techniques. Later Rodin and Warchawski [RWr '86] showed how to derive the sufficiency of the integral condition from known classical results, and T.F. Carroll [Crl '88] gave a potential theoretic proof of its necessity that generalizes to higher (real) dimensions. All of these results, including the original one of Tsuji and Warchawski, apply with slightly more complicated hypotheses to univalent maps that do not necessarily take the half-plane (or disc) into itself. Roughly speaking, the condition that the domain Ω lie in the upper half-plane can be replaced by a *minorant* condition entirely similar to the majorant condition discussed above.

5
Angular Derivatives and Iteration

In this chapter we are going to study the *dynamical behavior* of holomorphic self-maps of U. More precisely, if φ is such a map, then what can be said about its sequence of *iterates:*

$$\varphi_n = \varphi \circ \varphi \circ \cdots \circ \varphi \quad (n \text{ times})?$$

The connection between iteration and composition operators comes from the equation $C_{\varphi_n} = C_\varphi^n$, and this forms the basis of the work on cyclicity that takes place in Chapters 7 and 8. Right now, however, we intend to emphasize function theory, employing the methods developed for our theory of the angular derivative to study the dynamics of φ. As you might guess, our work will spin off consequences for the compactness problem.

5.1   Statement of Results

To get a feeling for what is going to happen here, suppose first that φ is a holomorphic self-map of U with a fixed point $p \in U$. There are two possibilities:

(a) *φ is an automorphism,* in which case $|\varphi'(p)| = 1$, so φ is elliptic, hence conjugate to the mapping of rotation through $\arg \varphi'(p)$. In this case an elementary exercise determines the behavior of the iterates of φ (see Exercise 1 below).

(b) *φ is not an automorphism,* in which case $|\varphi'(p)| < 1$ and, as we will see in the next section, the Schwarz Lemma shows that the iterates of φ converge to p.

Therefore, most of our work will focus on maps φ with no fixed point in U. For these we will prove the following remarkable theorem.

The Denjoy-Wolff Theorem. *If φ is a holomorphic self-map of U with no fixed point in U then there is a point $\omega = \omega(\varphi) \in \partial U$ such that $\varphi_n \to \omega$ uniformly on compact subsets of U.*

An elementary case of this result occurred in Chapter 0, where we observed that a linear fractional self-map of U with no interior fixed point has to be either hyperbolic or parabolic, with its attractive fixed point on ∂U. Beyond this, however, there is nothing obvious about the Denjoy-Wolff Theorem. As was the case for Littlewood's Subordination Principle, the subtlety is that no extra hypotheses are being assumed for φ—no extra valence restrictions, no extra regularity at the boundary. In this generality it is not clear how the magic "Denjoy-Wolff point" $\omega(\varphi)$ arises, or how the orbits of φ know where to find it.

These issues will be explained below, where the Denjoy-Wolff point will emerge as the unique "boundary fixed point" of φ at which the angular derivative φ exists and is ≤ 1. Thus, a map with no interior fixed point has, at its Denjoy-Wolff boundary point, all the extra regularity granted by the existence of the angular derivative, and that point acts very much like an interior fixed point for the map.

To be able to discuss this situation with some fluency, we declare a point $\omega \in \partial U$ to be a *boundary fixed point* of φ if $\varphi(r\omega) \to \omega$ as $r \to 1-$ (i.e. $\varphi(\omega) = \omega$ in the sense of radial limits). As in previous work we use the notation "$\overset{\kappa}{\to}$" to denote uniform convergence on compact subsets of U. Here is a precise statement that contains the Denjoy-Wolff Theorem, and summarizes most of what we will prove in this chapter. To appreciate parts (b) and (c), recall that the Julia-Carathéodory Theorem guarantees that if ω is a boundary fixed point of U at which φ has an angular derivative, then $\varphi'(\omega) > 0$.

The Grand Iteration Theorem. *Suppose φ is a holomorphic self-map of U that is not an elliptic automorphism.*

(a) *If φ has a fixed point $p \in U$ then $|\varphi'(p)| < 1$ and $\varphi_n \overset{\kappa}{\to} p$.*

(b) *If φ has no fixed point in U, then there is a point $\omega \in \partial U$ (the "Denjoy-Wolff point of φ") such that $\varphi_n \overset{\kappa}{\to} \omega$. Furthermore:*

   (i) *ω is a boundary fixed point of φ.*

   (ii) *The angular derivative of φ exists at ω, with $\varphi'(\omega) \leq 1$.*

(c) *Conversely, if φ has a boundary fixed point ω at which $\varphi'(\omega) \leq 1$, then φ has no fixed point in U, and ω is the Denjoy-Wolff point of φ.*

(d) *If $\omega \in \partial U$ is the Denjoy-Wolff point of φ, and $\varphi'(\omega) < 1$, then for each $z \in U$ the orbit $\{\varphi_n(z)\}$ converges nontangentially to ω.*

5.2 Elementary Cases

We begin with part (a) of the Grand Iteration Theorem, restated below for the convenience of the reader.

Proposition 1 (An interior fixed point). *Suppose φ is a holomorphic self-map of U that fixes a point p of U, and is not a conformal automorphism. Then $\varphi_n \xrightarrow{\kappa} p$.*

Proof. The key player in this argument is the ordinary Schwarz Lemma. Suppose first that $p = 0$, so $\varphi(0) = 0$. Then φ is not a rotation, so the Schwarz Lemma asserts that $|\varphi(z)| < |z|$ for every $z \in U$. This by itself is not enough to prove the desired convergence, but with a little more care it wins the day.

Fix $0 < r < 1$ and let $M(r)$ be the maximum of $|\varphi(z)|$ for $|z| \leq r$. By the Schwarz Lemma, $\delta \overset{\text{def}}{=} M(r)/r < 1$. The trick is to apply the Schwarz Lemma again, this time to the function

$$\psi(z) = \frac{\varphi(rz)}{M(r)} \qquad (z \in U).$$

The result is: $|\psi(z)| \leq |z|$ for all $z \in U$. Since ψ is continuous in the closed unit disc, there is a corresponding non-strict inequality there. The consequence for φ is that

$$|\varphi(z)| \leq \frac{M(r)}{r} |z| = \delta|z|$$

for each $z \in r\overline{U}$. Iteration of the last inequality yields:

$$|\varphi_n(z)| \leq \delta|\varphi_{n-1}(z)| \leq \delta^2|\varphi_{n-2}(z)| \leq \cdots \leq \delta^n|z| \leq \delta^n r$$

for each $z \in r\overline{U}$. Since $\delta < 1$, we see that $\varphi_n \to 0$ uniformly on $r\overline{U}$.

In case φ fixes a point $p \neq 0$, we need only apply the last result to the new function $\psi = \alpha_p \circ \psi \circ \alpha_p$, where α_p is the special (self-inverse) automorphism that interchanges p and 0 (see §0.4). The resulting holomorphic function ψ maps U into itself, fixes the origin, and is not an automorphism, so by the previous argument its iterates converge to 0 uniformly on compact subsets of U.

Now $\varphi = \alpha_p \circ \psi \circ \alpha_p$, and one checks easily that φ_n is given by the same formula, with ψ_n in place of ψ. Thus, the convergence of ψ_n to zero implies convergence of φ_n to p. This finishes the interior fixed point case of the iteration theorem. □

The next result is the direct "boundary analogue" of the result we have just proved; it is part (d) of the Grand Iteration Theorem.

Proposition 2 (A strongly attracting boundary fixed point.) *Suppose φ is a holomorphic self-map of U has an angular derivative at a boundary fixed point ω, and that $\varphi'(\omega) < 1$. Then for every $z \in U$ the orbit $\{\varphi_n(z)\}$ converges non-tangentially to ω.*

Proof. We may suppose without loss of generality that $\omega = +1$. We are asserting that for each $z \in U$ the orbit $\{\varphi_n(z)\}$ not only converges to $+1$, but also lies entirely inside one of the lens-shaped region L_α depicted in Figure 2.1 of §2.3.

Just as for the Julia-Carathéodory Theorem, the proof is best carried out in the right half-plane Π, where the transition from the disc is effected by the map $w = (1 + z)/(1 - z)$ that takes $+1$ to ∞, and replaces φ by $\Phi : \Pi \to \Pi$. We saw in the proof of the Julia-Carathéodory Theorem (see Remark at the end of §4.5) that whenever the angular derivative $\delta = \varphi'(1)$ exists, then

$$\operatorname{Re} \Phi(w) \geq \frac{1}{\delta} \operatorname{Re} w \qquad (w \in \Pi).$$

We are assuming that $\delta < 1$, so upon substituting $\Phi_n(w)$ for w in the inequality above, and iterating, we obtain

$$\operatorname{Re} \Phi_n(w) \geq \frac{1}{\delta^n} \operatorname{Re} w \to \infty,$$

which shows that the original orbit $\{\varphi_n(z)\}$ converges to $+1$, and identifies the real issue as the non-tangential nature of this convergence.

Fix $w \in \Pi$. We are going to show that the orbit $\{\Phi_n(w)\}$ lies in the half-plane image of some lens L_α, that is, in a sector

$$S_\alpha = \{w \in \Pi : |\arg w| < \alpha\}.$$

To simplify notation, let $w_n = \Phi_n(w)$, and write $\rho = d(w, w_1)$, the pseudo-hyperbolic distance between the first two points of the orbit. By the Invariant Schwarz Lemma (§4.3) we have $d(w_n, w_{n+1}) \leq \rho$ for each n, so using the notation of §4.3 (with $\overline{\Delta}$ denoting a *closed* pseudo-hyperbolic disc),

$$w_{n+1} \in \overline{\Delta}(w_n, \rho) = \operatorname{Re} w_n \, \overline{\Delta}(1, \rho) + i \operatorname{Im} w_n,$$

Figure 5.1 below, which is the analogue for this chapter of Figure 4.3 of §4.5, now tells the whole story.

The point is that w_{n+1} lies in the shaded region of $\overline{\Delta}(w_n, \rho)$, and this region is, in turn, the image of the corresponding shaded region of $\overline{\Delta}(1, \rho)$ under the affine map $A_n(w) = (\operatorname{Re} w_n)w + i \operatorname{Im} w_n$. Thus, $w_{n+1} \in w_n + S_\alpha$ for all n, which shows by induction that the whole orbit $\{w_n\}$ lies in the translated sector $w + S_\alpha$, and therefore in some sector having vertex at the origin, and an appropriately larger angular opening. $\qquad\square$

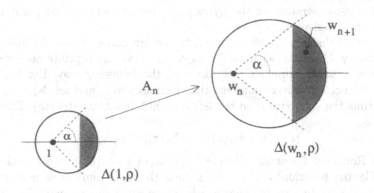

FIGURE 5.1. *Successive iterates* w_n *and* w_{n+1}

5.3 Wolff's Boundary Schwarz Lemma

The results of the last section show that if φ has an an angular derivative
at some boundary fixed point ω, and $\varphi'(\omega) < 1$, then φ cannot have an
interior fixed point, and ω is the Denjoy-Wolff point. Now suppose we only
know that φ has no interior fixed point. Why should we suspect that the
angular derivative even exists at some point of the boundary?

The answer lies in a beautiful geometric theorem due to Wolff, which can
be interpreted as a direct analogue of the Schwarz Lemma, where the role
of the fixed point at the origin is taken over by a point on the unit circle.

Wolff's Theorem. *Suppose φ is a holomorphic self-map of U that has no
fixed point in U. Then there is a unique point $\omega \in \partial U$ such that:*

(a) $\varphi(H) \subset H$ for any horodisc H in U tangent to ∂U at ω,

(b) ω is a boundary fixed point of φ, and

(c) φ has an angular derivative at ω, with $\varphi'(\omega) \leq 1$.

Proof. Not surprisingly, the "Wolff point" ω promised by this theorem will
eventually be revealed as the Denjoy-Wolff attracting point, from which will
follow its uniqueness. But we can prove right now that its uniqueness follows
from its existence! Indeed, suppose ω and ω' are Wolff points. Choose closed
horodiscs H and H' in U, with H tangent to ∂U at ω, and H' tangent to
∂U at ω', and further arrange these discs so they are tangent to each other
at some point $z_0 \in U$. Since φ is continuous on U, and ω and ω' are Wolff

points, the closure of each of these horodiscs gets taken into itself by φ. But z_0 is the lone point in both horodiscs, and it has to *stay* in both. Thus $\varphi(z_0) = z_0$, contradicting the hypothesis that φ has no fixed point in U. Thus, the Wolff point is unique.

To prove that the Wolff point exists, we are going to use the Julia-Carathéodory Theorem, with $\{p_n\}$ chosen to be an appropriate sequence of "approximate fixed points," obtained in the following way. Let $\{\rho_n\}$ be any sequence of positive numbers that increases to 1, and set $\Phi_n = \rho_n \varphi(z)$ (this time the subscript n on the left does not denote an iterate). Then for $|z| = \rho_n$,

$$|z - (z - \Phi_n(z))| = |\Phi_n(z)| < \rho_n = |z|$$

so by Rouché's Theorem ([Rdn '87, Theorem 10.43(b)]), the functions z (the identity function) and $z - \Phi_n(z)$ have the same number of zeros in the disc $\rho_n U$, namely *one*. In other words, there is a point p_n, with $|p_n| < \rho_n$, such that $\Phi_n(p_n) = p_n$. Thus p_n has the property

$$\varphi(p_n) = \frac{p_n}{\rho_n} \tag{1}$$

These are the "approximate fixed points" we seek.

Now pass to a subsequence, if necessary, to insure that the sequence of approximate fixed points converges to some point ω in the closed unit disc. Then by (1) the same is true of the image sequence: $\varphi(p_n) = p_n/\rho_n \to \omega$.

We claim that $\omega \in \partial U$. If not, $\omega \in U$, and the continuity of φ, along with (1), shows that $\omega = \lim \varphi(p_n) = \varphi(\omega)$, hence ω is a fixed point of φ in U; a contradiction.

To finish the proof of Wolff's Theorem, note that for each n, (1) implies that $|p_n| < |\varphi(p_n)|$, so

$$\frac{1 - |\varphi(p_n)|}{1 - |p_n|} < 1.$$

Thus by passing to a further subsequence if necessary, we can assume that the limit

$$\delta = \lim_n \frac{1 - |\varphi(p_n)|}{1 - |p_n|} \tag{2}$$

exists, and is ≤ 1. Julia's Theorem (§4.4) now asserts that φ takes every horodisc at ω into itself and that ω is a boundary fixed point of φ. Thus by (2) above and the Julia-Carathéodory Theorem, φ has an angular derivative at ω, and $\varphi'(\omega) \leq 1$. $\square$

Remark. If φ is not an automorphism, then the existence of a boundary point ω at which condition (a) of Wolff's Theorem holds is necessary and sufficient for φ to have no fixed point in U. Wolff's Theorem asserts the necessity. As for sufficiency, suppose such an ω exists, and, for the sake of contradiction, suppose that φ has a fixed point $p \in U$. Let H be a horodisc at ω that does not contain p, and choose $q \in H$. By hypothesis, the orbit of q has to stay in H, but by Proposition 1 of §5.2, this orbit should be converging to p. This is the desired contradiction.   □

5.4   Contraction Mappings

So far we know that if φ fixes no point of U then it has a boundary fixed point at which the angular derivative exists, and is ≤ 1. In case this derivative is < 1, we also know that the iterates of φ converge to ω. But what if $\varphi'(\omega) = 1$? We know from Wolff's Theorem that each orbit cannot get outside the horodisc at ω in which it started, but we have yet to isolate the cosmic agent that forces the orbits to converge to the boundary. This turns out to be nothing more than the Schwarz Lemma, dressed up as the fact that non-automorphisms strictly contract the pseudo-hyperbolic distance.

The Contraction Mapping Principle. *If φ is a holomorphic self-map of U that is not a conformal automorphism and has no fixed point in U, then $|\varphi_n| \to 1$ uniformly on compact subsets of U.*

Proof. It suffices to show that $|\varphi_n(0)| \to 1$. Suppose we have done this. Let K be a compact subset of U. The Invariant Schwarz Lemma asserts that φ shrinks the pseudo-hyperbolic distance between every pair of points, and from this one checks easily that

$$d(\varphi_n(K), \varphi_n(0)) \leq d(K, 0) < 1 \qquad (3)$$

for all n. Here the pseudo-hyperbolic distance from a point to a set is measured in the usual way: the infimum (minimum, in this case) of all the distances from the given point to points of the set. Thus $\varphi_n(K)$ has to head out to ∂U, in the sense that it eventually leaves any pre-assigned compact subset of U. For if this were not true, then by property (d) of the pseudo-hyperbolic distance in §4.3, the left-hand side of inequality (3) would have to tend to 1, which it does not.

To show that $|\varphi_n(0)| \to 1$, suppose otherwise, so there is a subsequence $\{n(k)\}$, and a point $p \in U$ such that $\varphi_{n(k)}(0) \to p$ as $k \to \infty$. To save notation, write $\varphi_n(0) = z_n$,

Now the Invariant Schwarz Lemma guarantees that the sequence of distances $\{d(z_n, z_{n+1})\}$ is decreasing, and therefore convergent to some $0 \leq \delta < 1$. By the continuity of both φ and the pseudo-hyperbolic distance,

$$d(p, \varphi(p)) = \lim_{k \to \infty} d(z_{n(k)}, z_{n(k)+1}) = \delta.$$

The pseudo-hyperbolic distance distinguishes points, so $\delta > 0$; otherwise φ would be a (forbidden) fixed point of φ in U. A repetition of the above limiting argument also shows

$$d(\varphi(p), \varphi_2(p)) = \lim_{k \to \infty} d(z_{n(k)+1}, z_{n(k)+2}) = \delta,$$

hence $d(p, \varphi(p)) = d(\varphi(p), \varphi_2(p)) > 0$, which contradicts the strict contractive property of φ. Thus $|z_n| \to 1$.   □

5.5   Grand Iteration Theorem, Completed

We can now complete the proof of the Grand Iteration Theorem. Only parts (b) and (c) remain.

Proof of part (b). Suppose φ fixes no point of U. Wolff's Theorem asserts that:

- There is a boundary fixed point ω such that $\varphi(H) \subset H$ for every horodisc H at ω, and

- the angular derivative of φ exists at ω, and has value ≤ 1.

So it remains only to show that $\varphi_n \overset{\kappa}{\to} \omega$.

Fix a compact subset K of U and choose a horodisc H at ω that contains K. By the Contraction Mapping Principle the sets $\varphi_n(K)$ converge to ∂U, but these sets are also confined to H, so they get to the boundary only by converging to ω.

Proof of part (c). Suppose ω is a boundary fixed point at which $\varphi'(\omega) \le 1$. Then

$$\limsup_{r \to 1-} \frac{1 - |\varphi(r\omega)|}{1 - r} \le 1$$

so by Julia's Theorem, $\varphi(H) \subset H$ for every horodisc H at ω. By the Remark at the end of the last section, φ therefore has no fixed point in U, hence the stage is set for the argument of the last paragraph to take over and finish the proof. The reader can easily supply the details.   □

Remark. It is possible for the map φ to have a myriad of boundary fixed points, at some of which there may even be a finite angular derivative (see Exercises 9 and 10 below). The above result asserts that φ has an angular derivative ≤ 1 at *no more than one* of these fixed points.

We close this chapter with an important application of the theory of iteration to the compactness problem for composition operators.

Corollary. *If the composition operator C_φ is compact on H^2, then φ has a fixed point in U.*

Proof. If φ has no fixed point in U, then by the result above, φ has an angular derivative at some point of ∂U. According to the Angular Derivative Criterion, C_φ is therefore not compact.   □

5.6 Exercises

1. *Elliptic automorphisms.* Fix $\alpha \in (-\pi, \pi]$ and consider the rotation map $T_\alpha : \partial U \to \partial U$ defined by $T_\alpha \omega = \alpha \omega$. Show that each orbit $\{T_\alpha^n \omega\}_0^\infty$ is:

   (a) A finite set if α is a rational multiple of π.

   (b) A dense subset of ∂U otherwise.

   Apply this result to determine the behavior of iterates of elliptic automorphisms.

2. Give a direct verification of the Denjoy-Wolff Theorem for linear fractional self-maps of U that have no fixed point in U.

3. *"Denjoy-Wolff for an interval."* Show that if f is a fixed-point-free homeomorphism of the open unit interval, then every f-orbit converges to an endpoint.

4. Construct a C^∞ (non-holomorphic) self-mapping of U that does not obey the conclusion of the Denjoy-Wolff Theorem.

5. Give an example of a holomorphic self-map of U with angular derivative $= 1$ at a boundary fixed point, for which every orbit converges to that fixed point non-tangentially. Give another example for which every orbit converges tangentially.

6. Suppose φ is a holomorphic self-map of U with a boundary fixed point ω at which $\varphi'(\omega) < 1$. Show that $\sum(1 - |\varphi_n(z)|) < \infty$ for every $z \in U$.

7. Suppose φ and ψ are holomorphic self-maps of U, and $\omega \in \partial U$. Show that if ω is the Denjoy-Wolff point for both φ and ψ, then it is the Denjoy-Wolff point for $\varphi \circ \psi$.

8. Continuing the problem above:

   (a) Suppose $\omega \in \partial U$ is the Denjoy-Wolff point of φ, and also a boundary fixed point of ψ. Characterize the maps ψ for which ω is the Denjoy-Wolff point of $\varphi \circ \psi$.

   (b) Write down some linear fractional examples to show that ω need not be the Denjoy-Wolff point of $\varphi \circ \psi$.

9. Give examples of holomorphic self-maps of U with:

(a) One boundary fixed point at which the angular derivative is finite, but > 1.

(b) Two boundary fixed points, neither of which is attractive. Is it possible for the angular derivative to be finite at both these fixed points?

10. *How to construct boundary fixed points.* Let G be the unbounded simply connected region bounded by the hyperbolas $y^2 - 4x^2 = 4$ and $x^2 - 4y^2 = 4$. Let σ be the Riemann mapping function that takes U onto G (so $\sigma(0) = 0$ and $\sigma'(0) > 0$). Define $\varphi(z) = \sigma^{-1}(\sigma(z)/2)$. Show that φ is a holomorphic self-map of U that has four fixed points on ∂U, none of which is attractive. Generalize. Can you use this idea to get *infinitely* many fixed points?

Suggestion: It's best to think about this problem on an intuitive level, believing what seems obvious. In particular: σ extends continuously to ∂U (where you regard ∂G as a subset of the Riemann Sphere), and the unit circle gets divided into four segments, each of which is mapped onto a different hyperbola branch. The map φ extends continuously to ∂U, and has the endpoints of these segments as fixed points.

We will discover in Chapter 9 that φ has an angular derivative at each of these fixed points.

5.7 Notes

Denjoy and Wolff independently and simultaneously proved the theorem that bears their name [Djy '26], [Wfl '26]. Soon afterward Wolff found a proof based on what we are calling Wolff's Theorem. The argument that uses Wolff's Theorem and the Contraction Mapping Principle appears in both [Brd '90] and [GlR '84, §30], and as both these sources show, it generalizes significantly to other situations that involve higher complex dimensions and complete Riemannian manifolds. For example, the "contraction–horodisc" argument given here extends almost without change to prove the Denjoy-Wolff Theorem for non-automorphic self-maps of the unit ball of $\mathbf{C}^n$, and even Hilbert space (the only significant change involves use of a fixed point theorem to produce the approximate fixed points required by the proof of Wolff's Theorem). [Brd '90] contains further valuable references to the literature, both classical and modern.

In surprising contrast to the situation for the unit disc, if one tries to prove the Denjoy-Wolff Theorem for the unit ball of $\mathbf{C}^n$, then analysis of automorphisms that fix just one point of $\mathbf{C}^n \cup \{\infty\}$ is no longer routine.

However, in this setting the Denjoy-Wolff Theorem survives; it was first proved by Hervé [Hrv '63], and versions of it were rediscovered by several authors about twenty years later (see [Kba '83], [Mlr '83] and [Ulr '83]).

For another approach to the part of the proof that follows from Wolff's Theorem, see Burckel's exposition [Brk '81], which also contains more detailed historical references, and further applications. For a survey of recent work in higher complex dimensions, see [Abt '91].

H.S. Bear [Br1 '92] [Br2 '92] has recently studied various aspects of the Denjoy-Wolff and Julia-Carathéodory theorems for distance decreasing maps of the hyperbolic plane, and Ky Fan has given extensions of the theorems of both Julia and Wolff to the setting of holomorphic functions acting in the sense of functional calculus on the unit ball of operators on Hilbert space [Fan '78, '83].

6
Compactness and Eigenfunctions

In previous chapters we explored the connection between compactness for composition operators and the existence of angular derivatives for their inducing maps. Here we begin to study how compactness affects the *eigenfunctions* of a composition operator. The eigenfunction equation for a composition operator C_φ is called *Schröder's equation*:

$$f \circ \varphi = \lambda f, \tag{1}$$

and has been studied in one form or another since the late nineteenth century [Shr '71].

In the last chapter, we discovered as a consequence of the Julia-Carathéodory and Denjoy-Wolff theorems that whenever a composition operator is compact on H^2, its inducing map must have a fixed point in the unit disc. Königs in 1884 initiated the study of Schröder's equation for holomorphic self-maps of the disc having an interior fixed point, and found the eigenvalues λ for which holomorphic solutions exist [Kgs '84]. We begin this chapter by reviewing Königs's work, and then carry it a step further by showing that compactness of C_φ on H^2 results in a significant growth restriction on the solutions of Schröder's equation. This work mixes function theory with functional analysis, and also leads to the determination of the *spectrum* of a compact composition operator. What is more important, our study of eigenfunctions leads naturally to the idea of representing univalent self-maps of U by linear fractional maps that act on more complicated domains, where the geometry of the domain encapsulates the subtleties of the original map. Such "linear fractional models" will, in turn, play a large role in Chapters 8 and 9.

6.1 Königs's Theorem

In this section we give Königs's solution of Schröder's equation for holomorphic maps φ with an interior fixed point. The work will take place in $H(U)$, the full space of holomorphic functions on U, with Hilbert space issues set aside until the next section.

Our discussion of Königs's Theorem proceeds in two stages; first we list the properties that solutions must possess, and prove that each solution has "multiplicity one." Then we prove that "all solutions that could possibly exist actually do exist."

Proposition (Necessity and uniqueness). *Suppose φ is a holomorphic self-map of U that fixes a point $p \in U$ and is neither a constant nor a conformal automorphism of the unit disc. Suppose $f \in H(U)$ is not constant, and there is a complex number λ such that f and λ satisfy Schröder's equation (1). Then*

(a) *$\varphi'(p) \neq 0$ and $\lambda = \varphi'(p)^n$ for some positive integer n.*

(b) *Up to a multiplicative constant, f is the unique solution for $\lambda = \varphi'(p)^n$.*

(c) *$f(p) = 0$.*

Proof. First, let us observe that λ is neither 0 nor 1. For if $\lambda = 0$ then Schröder's equation becomes $f \circ \varphi = 0$, so f vanishes on $\varphi(U)$ which is a non-void open set because φ is holomorphic and non-constant. Thus f vanishes identically on U, which contradicts the assumption that f is not constant. So $\lambda \neq 0$. If $\lambda = 1$ then f satisfies the equation $f \circ \varphi = f$, so more generally $f \circ \varphi_n = f$ for each positive integer n. But we saw in §5.2 that for each $z \in U$ the orbit $\{\varphi_n(z)\}$ must converge to the interior fixed point, hence

$$f(z) = \lim_n f(\varphi_n(p)) = f(p).$$

Thus only constant functions can satisfy Schröder's equation with $\lambda = 1$ (and note that for $\lambda = 1$ every constant function *does* satisfy the equation). Note also that Schröder's equation implies that

$$f(p) = f(\varphi(p)) = \lambda f(p),$$

so knowing now that $\lambda \neq 1$ we conclude that $f(p) = 0$.

For the rest of the proof, and indeed for the rest of the chapter, we assume that $p = 0$. This entails no loss of generality, since if it is not already the case, then the map $\psi = \alpha_p \circ \varphi \circ \alpha_p$ has fixed point at the origin, and for a given complex number λ, solutions of Schröder's equation for φ and ψ are related by the equation $f = F \circ \alpha_p$. We leave it to the reader to check, each time a proof is over, that all essential properties are preserved upon conjugation by a conformal automorphism.

To identify λ, note that since f is not constant, and vanishes at the origin, there exists a positive integer n such that

$$f(z) = a_n z^n + a_{n+1} z^{n+1} + \cdots,$$

where $a_n \neq 0$. Upon solving Schröder's equation for λ we obtain for each $z \in U$:

$$\lambda = \frac{f(\varphi(z))}{f(z)} = \left(\frac{\varphi(z)}{z}\right)^n \frac{a_n + a_{n+1}\varphi(z) + a_{n+2}\varphi(z)^2 + \cdots}{a_n + a_{n+1}z + a_{n+2}z^2 + \cdots},$$

and it is clear that the right-hand side of this equation converges to $\varphi'(0)^n$ as $z \to 0$.

Only the uniqueness of f remains to be proved. Upon successively differentiating both sides of Schröder's equation, and evaluating each result at $z = 0$, we find that for each $n \geq 2$, the quantity $(\lambda - \lambda^n)f^{(n)}(0)$ is given by an expression that involves the derivatives $\varphi^{(k)}(0)$ for $1 \leq k \leq n$, and $f^{(j)}(0)$ for $1 \leq j \leq n-1$. Since λ is neither 0 nor 1, an induction argument shows that $f^{(n)}$ is determined solely by φ and $f'(0)$. $\square$

At this point we know that the eigenvalues of Schröder's equation (if indeed there are any) lie among the numbers $\{\varphi'(0)^n\}$. But does every number $\varphi'(0)^n$ correspond to a non-trivial solution? Königs showed that if $0 < |\varphi'(0)| < 1$ then there is a solution σ for $\varphi'(0)$. Once this is done, σ^n is clearly the (essentially unique) solution for $\varphi'(0)^n$.

Proposition (Existence). *Suppose φ is a holomorphic self-map of U that fixes a point $p \in U$, is not an (elliptic) automorphism of U, and has non-vanishing derivative at p. Then there exists $\sigma \in H(U)$ such that $\sigma \circ \varphi = \varphi'(0)\sigma$. Moreover, if φ is univalent, then so is σ.*

Proof. The function σ will be obtained as the limit of a sequence of "normalized iterates" of φ. As previously discussed, we may take $p = 0$, so that $\varphi(0) = 0$. By the Schwarz Lemma, $0 < |\varphi'(0)| < 1$. Let $\lambda = \varphi'(0)$, and for each positive integer n set $\sigma_n = \lambda^{-n}\varphi_n$ (noting that while the subscript n on the right-hand side of the equation denotes an iterate, the one on the left does not). Since $\sigma_n \circ \varphi = \lambda\sigma_{n+1}$ for each n, the solution σ we seek will be the κ-limit of the sequence of functions $\{\sigma_n\}$, if only we can prove that this limit exists.

To do so, write

$$\sigma_n(z) = z \cdot \frac{\varphi(z)}{\lambda z} \cdot \frac{\varphi_2(z)}{\lambda \varphi(z)} \cdot \ldots \cdot \frac{\varphi_n(z)}{\lambda \varphi_{n-1}(z)} = z \prod_{j=0}^{n-1} F(\varphi_j(z)),$$

where in the product on the right, $\varphi_0(z) \equiv z$, and $F(z) = \varphi(z)/\lambda z$. Thus our task is to prove that the infinite product

$$\prod_{j=0}^{\infty} F(\varphi_j(z))$$

converges uniformly on compact subsets of U. For this it is enough to show that the infinite series

$$\sum_{j=0}^{\infty} |1 - F(\varphi_j(z))| \tag{2}$$

converges likewise. We estimate the size of each term. Since $\varphi(0) = 0$, the function F is holomorphic on U. By the Maximum Principle, $\|F\|_\infty \leq |\lambda^{-1}|$, hence

$$\|1 - F\|_\infty \leq 1 + \|F\|_\infty \leq 1 + \frac{1}{|\lambda|} \overset{\text{def}}{=} A.$$

Finally, the definition $\lambda = \varphi'(0)$ forces $F(0) = 1$, so the Schwarz Lemma, applied to the function $(1 - F)/A$, yields

$$|1 - F(z)| \leq A|z| \qquad (z \in U). \tag{3}$$

Now fix $0 < r < 1$. In the work of §5.2 we observed that a slight refinement of the Schwarz Lemma produces a constant $M(r) < 1$ such that for each non-negative integer j,

$$|\varphi_j(z)| \leq M(r)^j |z|$$

whenever $|z| \leq r$. Upon substituting this inequality in (3) we see that

$$|1 - F(\varphi_j(z))| \leq A|\varphi_j(z)| \leq AM(r)^j |z| \qquad (z \in r\overline{U}).$$

So on the closed disc $r\overline{U}$, each term of the series (2) is bounded uniformly by the corresponding term of a convergent geometric series, hence the original series converges uniformly on that disc. This establishes the desired convergence for the sequence $\{\sigma_n\}$.

Finally, If φ is univalent, then so is each normalized iterate σ_n. Since $\sigma_n \overset{\kappa}{\to} \sigma$, Hurwitz's Theorem ([Rdn '87], Chapter 10, Problem 20), implies that σ is either univalent or constant. But σ is not constant since each σ_n is defined to have derivative $= 1$ at the origin, hence σ has the same property.   □

From now on we reserve the symbol σ to denote the unique eigenfunction of Schröder's equation for $\lambda = \varphi'(p)$ that has $\sigma'(p) = 1$. This function is called the *Königs function* of φ, or the *principal eigenfunction* of C_φ. If $p = 0$ then σ is the function produced above as a limit of normalized iterates of φ. In the general case it is obtained from the case $p = 0$ by conjugation with the automorphism α_p.

Here is a summary of what we have just done.

Königs's Theorem (1884). *Suppose φ is a non-constant, non-automorphic holomorphic self-map of U with a fixed point $p \in U$, and consider C_φ as a linear transformation of $H(U)$. Then*

(a) *If $\varphi'(p) = 0$, then $+1$ is the only eigenvalue of C_φ.*

(b) *If $\varphi'(0) \neq 0$ then the eigenvalues of C_φ are precisely the numbers $\{\varphi'(p)^n\}_0^\infty$. Each has multiplicity one, and the function σ^n spans the eigenspace for $\varphi'(p)^n$ $(n = 1, 2, \ldots)$.*

(c) *If φ is univalent, then so is its Königs function σ.*

In case φ is univalent, the fact that the Königs function is also univalent invests Schröder's equation with important geometric significance. To understand what is at stake here, let $\lambda = \varphi'(0)$ (non-zero by the univalence of φ), and observe that Schröder's equation insures that:

(a) The simply connected domain $\sigma(U)$ is taken into itself upon multiplication by λ, and

(b) The map φ has the representation

$$\varphi(z) = \sigma^{-1}(\lambda \sigma(z)) \qquad (z \in U).$$

In other words,

> Upon conjugation by σ the mapping φ is equivalent to multiplication by λ, acting on $\sigma(U)$.

From this point of view, we understand φ by understanding the subtleties of σ, or equivalently, the geometry of $G = \sigma(U)$. This is the crucial idea in much of the work to follow. For example, the rest of this chapter, and all of Chapter 9, concerns the problem of how the geometry of G determines whether or not C_φ is compact. To put the issue more succinctly: suppose you draw a picture of a region G that contains the origin and is taken into itself under multiplication by (say) $1/2$. Let σ be a univalent mapping of U onto G with $\sigma(0) = 0$. Define φ by $\varphi(z) = \sigma^{-1}(\frac{1}{2}\sigma(z))$ for $z \in U$. How do you tell, by looking at G, if C_φ is compact?

6.2 Eigenfunctions for Compact C_φ

In this section we consider the problem of determining when the eigenvalues determined by Königs's Theorem are eigenvalues for the operator C_φ acting as an operator on H^2. In other words:

> When does σ^n belong to H^2 for every positive integer n?

A couple of examples show what can happen here, and at the same time illustrate the "multiplication mapping" point of view advanced at the end of the last section.

Example (No non-constant eigenfunctions in H^2). The mapping

$$\sigma(z) = \frac{1+z}{1-z} - 1 = \frac{2z}{1-z} = \sum_{k=1}^{\infty} 2z^k.$$

takes the unit disc conformally onto the half-plane $G = \{\operatorname{Re} z > -1\}$, and clearly does not belong to H^2. Let φ denote the holomorphic self-map of U that is conjugate via σ to the mapping of multiplication by $1/2$ on G, that is,

$$\varphi(z) = \sigma^{-1}\left(\frac{1}{2}\sigma(z)\right) = \frac{z}{2-z} \qquad (z \in U).$$

Then φ satisfies Schröder's equation with $\lambda = \varphi'(0) = 1/2$.

Our integral representations of the H^2 norm show that once a function is not in H^2, then neither is any positive integer power of that function. Thus for this example, *none* of the eigenfunctions above belong to H^2.

Example (All eigenfunctions in H^2). Recall the *lens maps* φ_α introduced in §2.3. For $0 < \alpha < 1$ the map φ_α was constructed by first mapping the unit disc conformally to the right half-plane, then taking the α-th power, and returning to the unit disc by the inverse of the half-plane mapping. However the α-th power map, acting on the right half-plane, can be represented, via the logarithm function, by multiplication by α on the strip $\{|\operatorname{Im} w| < \pi/2\}$. Putting it all together (and leaving the details to the reader),

$$\sigma(\varphi_\alpha(z)) = \alpha\,\sigma(z) \qquad (z \in U),$$

where

$$\sigma(z) = \log\frac{1+z}{1-z}$$

is the Königs function for each φ_α. We saw early in Chapter 1 that this particular σ belongs to H^2, and a little more work, using the integral representation of the H^2 norm given in either §1.2 or §3.1, shows that $\sigma^n \in H^2$ for every positive integer n.

Note that in the first of these examples, φ is a linear fractional map that takes the unit disc onto a horodisc tangent to the unit circle at the point $+1$, so by the Angular Derivative Criterion (§4.2) the induced composition operator is not compact. On the other hand, we also saw in Chapter 2 that the lens maps of the second example all induce compact composition operators. These examples raise the suspicion that there must be a connection between compactness for C_φ and restricted growth for the Königs function σ, or equivalently, restricted size for $\sigma(U)$. The next result, which is the goal of this section, shows that there is some merit to this idea.

The Eigenfunction Theorem. *If φ is a holomorphic self-map of U for which C_φ is compact on H^2, then $\sigma^n \in H^2$ for every positive integer n.*

Our proof of this theorem requires a powerful result of elementary operator theory. In order to state it efficiently, let us agree to call an operator on Hilbert space *invertible* if it maps the space one-to-one onto itself. The Open Mapping Theorem insures that the inverse of such an operator is automatically bounded.

Riesz's Theorem. *If T is a compact operator on a Hilbert space $\mathcal{H}$, then the following four conditions are equivalent:*

(a) $I - T$ is invertible on $\mathcal{H}$.

(b) $I - T$ is one-to-one on $\mathcal{H}$.

(c) $(I - T)(\mathcal{H}) = \mathcal{H}$.

(d) $I - T^*$ is invertible on $\mathcal{H}$.

Riesz's Theorem asserts that, relative to invertibility, compact perturbations of the identity behave just like finite dimensional operators. Recall that if T is compact then so is its adjoint T^* (§3.4), hence we can add to this list of equivalences versions of items (b) and (c) where T^* replaces T.

We defer the proof of Riesz's Theorem to the end of the chapter.

Proof of the Eigenfunction Theorem. Since C_φ is compact, we know from the last result of Chapter 5 that φ fixes a point of U, and as always, there is no loss of generality in assuming that this point is the origin.

Let $\mathcal{P}_n$ denote the subspace of H^2 that consists of polynomials of degree $\leq n$:

$$\mathcal{P}_n = \text{span } \{1, z, z^2, \ldots, z^n\}.$$

We claim that each subspace $\mathcal{P}_n$ is *invariant* under C_φ^*, i.e.,

$$C_\varphi^*(\mathcal{P}_n) \subset \mathcal{P}_n.$$

To see why this is true, observe that since the power series representation of φ has constant term zero, then the coefficients of index 0 through $m - 1$ all vanish in the corresponding representation for φ^m. If $m > n$ then it follows that every polynomial of degree $\leq n$ is orthogonal in H^2 to φ^m. If P is such a polynomial, then

$$< C_\varphi^*(P), z^m > = < P, C_\varphi(z^m) > = < P, \varphi^m > = 0.$$

Consequently

$$C_\varphi^*(P) \subset (\text{span } \{z^m : m > n\})^\perp = \mathcal{P}_n,$$

which establishes our claim.

The invariance of each of the subspaces $\mathcal{P}_n$ is equivalent to the fact that C_φ^* has an upper triangular matrix with respect to the orthonormal basis $\{z^n\}$ of monomials. Now the calculation

$$< C_\varphi^*(z^n), z^n > = < z^n, C_\varphi(z^n) > = < z^n, \varphi^n > = \overline{\varphi'(0)^n}$$

(n a non-negative integer) shows that the diagonal entries of this matrix are just the successive powers of $\overline{\varphi'(0)}$.

Now fix a non-negative integer n. The matrix of the restriction of C_φ^* to $\mathcal{P}_n$, with respect to the monomial basis $\{z^k\}_0^n$, is just the upper left $n \times n$ block of the big matrix. In particular, $\overline{\varphi'(0)}\,n$ is an eigenvalue of this restriction, and, therefore, an eigenvalue of the original operator. Now apply the Riesz Theorem (specifically, the implication (d) $\rightarrow$ (b)) to the operator $T = \varphi'(0)^{-n} C_\varphi$.

Conclusion: $I - T$ is not is not one-to-one, hence neither is

$$C_\varphi - \varphi'(0)^n I = \varphi'(0)^n (T - I).$$

Thus $\varphi'(0)^n$ is an eigenvalue of C_φ. But Königs's Theorem asserts that the corresponding eigenvector (which must belong to H^2, since that is the space on which C_φ is acting in this proof) must be a non-zero constant multiple of σ^n, where σ is the Königs function for φ. Thus, each positive integer power of the Königs function lies in H^2.   □

See Exercise 12 below for more on this proof.

Spectra. The *spectrum* of a Hilbert space operator T is the set of complex numbers λ for which the operator $T - \lambda I$ is not invertible. The Riesz Theorem implies that the non-zero spectral points of a compact operator are precisely the non-zero eigenvalues. Since no compact operator on an infinite dimensional Hilbert space is invertible (if it were, the unit ball would have to be relatively compact, and the space therefore finite dimensional), zero is also a point of the spectrum. Thus the spectrum of a compact operator is precisely its collection of eigenvalues, along with zero. The work of this section therefore provides the following description of the spectrum of a compact composition operator.

Corollary. *If C_φ is compact on H^2, and $p \in U$ is the fixed point of φ, then*

$$\text{Spectrum of } C_\varphi = \{0, 1\} \cup \{\varphi'(p)^n : n = 1, 2, \ldots\}.$$

6.3 Compactness vs. Growth of σ

The Eigenfunction Theorem can also be interpreted as a "slow growth" result on the Königs function σ. In what follows, φ is a holomorphic self-map of U with a fixed point in U, and σ is its Königs function.

Corollary. *If C_φ is compact on H^2, then for each $\varepsilon > 0$,*

$$|\sigma(z)| = O\left(\frac{1}{(1 - |z|)^\varepsilon}\right) \qquad (|z| \to 1-).$$

Proof. Fix a positive integer n. Since $\sigma^n \in H^2$, the Growth Condition of §1.1 shows that

$$|\sigma^n(z)| = O(1 - |z|)^{-1/2} \qquad (\text{as } |z| \to 1-),$$

and this yields the desired result. $\qquad\qquad\qquad\qquad\qquad\qquad\qquad$ $\square$

The same reasoning yields a somewhat more precise growth condition. For $0 < p < \infty$ let H^p be the space of functions $f \in H(U)$ for which

$$\sup_{0 \le r < 1} \frac{1}{2\pi} \int_{-\pi}^{\pi} |f(re^{i\theta})|^p d\theta < \infty.$$

As p increases, the L^p norms in this definition also increase, so the spaces H^p decrease in size. In terms of these spaces, the Eigenfunction Theorem can be rephrased as follows:

Theorem. *If C_φ is compact on H^2, then $\sigma \in H^p$ for every $p < \infty$.*

Recall that for each of the lens maps φ_α, the Königs function σ is a conformal mapping of U onto an infinite strip. Thus compactness for C_φ does not force σ to be bounded (see Exercise 1 below for conditions that guarantee σ is bounded).

It is also interesting to observe that the various sufficient conditions for compactness discovered in Chapters 2 through 4 combine with the Eigenfunction Theorem to give the following purely classical result about slow growth of the Königs function.

Corollary. *If either:*
(a) φ is univalent and has no angular derivative on ∂U, or

(b) $\displaystyle \int_{-\pi}^{\pi} \frac{1}{1 - |\varphi(e^{i\theta})|} \, d\theta < \infty,$

then the Königs function of φ belongs to H^p for every $p < \infty$.

6.4 Compactness vs. Size of $\sigma(U)$

The Eigenfunction Theorem has geometric consequences of the form "C_φ compact implies $\sigma(U)$ small." The following is a more precise statement.

Corollary. *Suppose φ is a univalent self-map of U that has an interior fixed point. Let σ be the Königs function of φ. If C_φ is compact on H^2, then $\sigma(U)$ contains no angular sector.*

Proof. We prove the contrapositive statement. Suppose $\sigma(U)$ contains a sector, i.e., that there exists a point $P \in \mathbf{C}$, a complex number $|\omega| = 1$, and an "aperture" $0 < \alpha < 1$ such that

$$\sigma(U) \supset S = P + \omega \cdot S_{\alpha\pi},$$

where $S_{\alpha\pi}$ is the sector with vertex at the origin that is symmetric about the positive real axis, and has angular opening $\alpha\pi$. Our goal is to show that C_φ is not compact.

We may assume that $\alpha = 1/n$ for some positive integer n. Thus, the map

$$\tau(z) = P + \omega \cdot \left(\frac{1+z}{1-z}\right)^{1/n},$$

takes U holomorphically onto S, and it is easy to see that

$$\left(\frac{\tau(z) - P}{\omega}\right)^n = \frac{1+z}{1-z} \notin H^2.$$

Thus $\tau^n \notin H^2$, and (repeating a now-familiar argument) $\tau = \sigma \circ \chi$ where $\chi = \sigma^{-1} \circ \tau$ is a holomorphic self-map of U. It follows that $\tau^n = \sigma^n \circ \chi$, hence $\sigma^n \notin H^2$ (otherwise Littlewood's Theorem (§1.3) would force $\tau^n \in H^2$). Thus, we conclude from the Eigenfunction Theorem that C_φ is not compact.   □

The Corollary may, of course, be restated as a result that links the angular derivative of φ with the geometry of its Königs domain:

> If p is a univalent self-map of U with an interior fixed point, and $\sigma(U)$ contains a sector, then φ has an angular derivative at some point of ∂U.

For an example of this result in action, see Exercise 4 below.

A more complicated argument, which we will not present, shows that C_φ will not be compact even if $\sigma(U)$ is only assumed to contain a "twisted sector" (see [SSS '92]). In Chapter 9 we show that if $\sigma(U)$ satisfies a natural starlikeness condition, then the Corollary can be turned around to give a "no sectors implies compactness" result. Thus, for a large class of natural mappings we will eventually be able to decide the compactness of the induced composition operator simply by looking at a picture of $\sigma(U)$.

6.5   Proof of Riesz's Theorem

We organize the argument into several lemmas. The standing notation is that T is a bounded linear operator on a Hilbert space $\mathcal{H}$, while $\ker T$ denotes the null space of T and $\operatorname{ran} T = T(\mathcal{H})$. We denote the closure of a subset S of $\mathcal{H}$ by $\overline{S}$.

Lemma 1. $\ker T^* = (\operatorname{ran} T)^\perp$, and $(\ker T)^\perp = \overline{\operatorname{ran}(T^*)}$.

Proof. Recall from §3.4 the defining equation for the adjoint operator:

$$< Tx, y >=< x, T^*y > \qquad (x, y \in \mathcal{H}). \qquad (4)$$

This shows that y is orthogonal to $\operatorname{ran} T$ if and only if T^*y is orthogonal to every $x \in \mathcal{H}$, i.e., $T^*y = 0$. This proves the first identity. For the second, replace T by T^* in the first, and use the fact that $T^{**} = T$ (obvious from (4) above). The result is: $\ker T = (\operatorname{ran} T^*)^\perp$. The second identity follows from this and the fact that the "double perp" of a subspace is just its closure (see [Rdn '87], Chapter 4, Problem 1). $\square$

For the rest of this section, T is a compact operator on $\mathcal{H}$, and $S = I - T$. We call S a *compact perturbation of the identity*.

Lemma 2. $\ker S$ is finite dimensional and $\operatorname{ran} S$ is closed.

Proof. Let $\mathcal{K} = \ker S$. The restriction of T to $\mathcal{K}$ is both the identity map and a compact operator on $\mathcal{K}$. Thus, the (closed) unit ball of $\mathcal{K}$ is compact, so $\mathcal{K}$ is finite dimensional.

To show $\operatorname{ran} S$ is closed, let y be one of its limit points, so $y = \lim S x_n$ for some sequence $\{x_n\}$ in $\mathcal{H}$. We want to show that $y = Sx$ for some $x \in \mathcal{H}$.

This is easy if the sequence $\{x_n\}$ is bounded, since now the compactness of T insures that the image sequence $\{T x_n\}$ has a subsequence that converges to some $z \in \mathcal{H}$. Upon discarding everything but this subsequence, we may assume that $z = \lim T x_n$. Then

$$\lim_n x_n = \lim_n (T x_n + S x_n) = z + y.$$

Writing $x = z + y$ and using the continuity S, we obtain $y = \lim S x_n = Sx$, as desired.

If $\{x_n\}$ is not bounded we can nevertheless replace it by a bounded sequence $\{\xi_n\}$ such that $S\xi_n = S x_n$ for each n. To see how this goes, take ξ_n to be the component of x_n that is orthogonal to $\mathcal{K}$. Indeed, $x_n = \xi_n + \eta_n$, where $\eta_n \in \mathcal{K}$ and $\xi_n \perp \mathcal{K}$, so clearly $S x_n = S\xi_n$. We claim that the sequence $\{\xi_n\}$ is bounded, in which case the theorem follows from the previous case.

So suppose, for the sake of contradiction, that $\{\xi_n\}$ is *not* bounded. By passing to an appropriate subsequence we may assume that $\|\xi_n\| \to \infty$. Let $e_n = \xi_n/\|\xi_n\|$. Then upon recalling that $Sx_n \to y$, and we have

$$Se_n = \frac{Sx_n}{\|\xi_n\|} \to 0.$$

Because T is compact, we obtain, after discarding everything but an appropriate subsequence,

$$Te_n \to \text{some } w \in \mathcal{H}.$$

Now w is not just any vector! On one hand,

$$e_n = Se_n + Te_n \to 0 + w = w, \tag{5}$$

so $w \perp \mathcal{K}$ (since each $e_n \perp \mathcal{K}$). But it follows from (5) that

$$0 = \lim_n Se_n = Sw,$$

so on the other hand, $w \in \mathcal{K}$. Thus $w = 0$, so (5) actually reads: $e_n \to 0$, and this contradicts the fact that each e_n is a unit vector.   □

These two lemmas immediately yield the following result.

Corollary. ran $S = (\ker S^*)^\perp$, *a closed subspace of finite codimension.*

This corollary tells us that any compact perturbation of the identity has closed range of finite codimension, and is onto if and only if its adjoint is one-to-one. In order to prove Riesz's Theorem we have to obtain this last conclusion for the operator itself, rather than its adjoint. To this end define for each non-negative integer n the "generalized eigenspaces" $\mathcal{K}_n = \ker S^n$. Thus

$$\{0\} = \mathcal{K}_0 \subset \mathcal{K}_1 \subset \mathcal{K}_2 \cdots . \tag{6}$$

The next result asserts that this sequence of subspaces "eventually stabilizes."

Lemma 3. *There exists an integer $n \geq 0$ such that the chain of inclusions (6) is strict up to index n, with equality thereafter.*

Proof. Note first that as soon as there is equality between two successive terms of (6), there is equality ever after. For suppose $\mathcal{K}_n = \mathcal{K}_{n+1}$ for some n. We need only show that $\mathcal{K}_{n+1} = \mathcal{K}_{n+2}$. For this, suppose $x \in \mathcal{K}_{n+2}$, so $0 = S^{n+2}x = S^{n+1}Sx$. Thus $Sx \in \mathcal{K}_{n+1} = \mathcal{K}_n$, so $S^{n+1}x = S^n(Sx) = 0$, which shows that $x \in \mathcal{K}_{n+1}$. Thus $\mathcal{K}_{n+2} \subset \mathcal{K}_{n+1}$, and since the opposite inclusion is always true, there is equality.

Therefore, it suffices to prove that the spaces $\mathcal{K}_n$ are not all distinct. Suppose this is not true, so there is strict inclusion throughout (6). Then for each n we can find a unit vector $x_n \in \mathcal{K}_{n+1}$ that is orthogonal to $\mathcal{K}_n$. We are going to show that the image sequence $\{Tx_n\}$ has no convergent subsequence, in contradiction to the assumed compactness of T. Once done, this will complete the proof of the Lemma.

Fix $n > m$ and observe that both Sx_n and Tx_m belong to $\mathcal{K}_n$ (the latter because T commutes with S). Thus the sum also belongs to $\mathcal{K}_n$, so after a little algebra,

$$Tx_n - Tx_m \in x_n + \mathcal{K}_n.$$

Since $x_n \perp \mathcal{K}_n$, we have

$$\|Tx_n - Tx_m\| \ge \|x_n\| = 1,$$

which establishes the claim, and proves the Lemma. $\square$

Lemma 4. $\ker S = \{0\} \iff \operatorname{ran} S = \mathcal{H}$.

Proof. Suppose that $\operatorname{ran} S = \mathcal{H}$. Then given any vector $x_0 \in \mathcal{H}$, and a positive integer n, we can find $x_n \in \mathcal{H}$ such that $S^n x_n = x_0$. Now if $\ker S \ne \{0\}$ we can choose $0 \ne x_0 \in \ker S$, from which it follows that for each positive integer n, the vector x_n belongs to $\mathcal{K}_{n+1}$, but not to $\mathcal{K}_n$. This contradicts Lemma 3. Thus $\ker S = \{0\}$.

Conversely, suppose $\ker S = \{0\}$. Then by the Corollary following Lemma 2, $\operatorname{ran} S^* = \mathcal{H}$. Now $S^* = I - T^*$, and T^* is also compact, so by the last paragraph, with T^* in place of T, we see that $\ker S^* = \{0\}$. By the Corollary and the fact that $S^{**} = S$, we conclude that $\operatorname{ran} S = \mathcal{H}$, as desired. $\square$

Lemma 4 finishes the proofs of parts (a) through (c) of Riesz's Theorem. Part (d) follows from the fact that the adjoint of a compact perturbation of the identity is another such operator. $\square$

6.6 Exercises

1. Suppose φ is a holomorphic self-map of U such that $\|\varphi_n\|_\infty < 1$ for some positive integer n (recall that φ_n denotes the n-th iterate of φ). Show that the Königs function of φ is bounded.

2. Prove the converse of the result of Exercise 1 for *univalent* maps φ.

3. *Local Königs's Theorem.* (a) Suppose f is a function holomorphic in a neighborhood of the origin, with $f(0) = 0$ and $0 < |f'(0)| < 1$. Show that Schröder's equation $\sigma \circ f = \lambda \sigma$ (where $\lambda = f'(0)$) for

f is solvable for some function σ that is holomorphic and univalent on a neighborhood of the origin, and uniquely determined by the normalization $\sigma'(0) = 1$.

(b) Suppose that f is as in part (a), except that $\lambda = f'(0)$ is a root of unity. Suppose that f is not the function $z \mapsto \lambda z$. Show that Schröder's equation for f has no solution σ holomorphic in a neighborhood of the origin.

4. Show that the univalent self-map of U defined in Exercise 10 of Chapter 5 has an angular derivative at each of its four boundary fixed points.

The next four problems show that non-compact composition operators, even those introduced by linear fractional maps, can have surpisingly diverse eigenvalue structure.

5. Let $\varphi(z) = (1 + z)/2$. Find all complex numbers α for which the function $(1 - z)^\alpha$ is an H^2-eigenfunction for the composition operator C_φ, and find the corresponding eigenvalues. Use your result to provide yet another proof that C_φ is not compact for this particular map φ.

6. Suppose φ is a parabolic automorphism of U. Show that the H^2-eigenvalues of C_φ are precisely the points of the unit circle, and that each has infinite multiplicity.

   Suggestion: Use a linear fractional map to take the unit disc onto the right half-plane, sending the fixed point to ∞. Then the map of the half-plane that corresponds to φ is translation by a pure imaginary number.

7. Suppose $\varphi \in LFT(U)$ is a parabolic non-automorphism of U. Show that the eigenvalues of φ form a curve in U that starts at the fixed point and approaches the origin, and that each eigenvalue has multiplicity one.

8. Suppose φ is a hyperbolic automorphism of U with attractive fixed point α. Show that the eigenvalues of φ are precisely the points of the annulus $\{|\varphi'(\alpha)|^{1/2} < |\lambda| < |\varphi'(\alpha)|^{-1/2}\}$, and that each eigenvalue has infinite multiplicity.

The next three problems concern the notion of similarity for bounded operators, and in particular for composition operators.

9. Hilbert space operators T_1 and T_2 are said to be *similar* if there is an invertible operator S such that $T_2 = S^{-1}T_1S$. Composition operators C_φ and C_ψ are said to be *compositionally similar* if there is a conformal automorphism α of the unit disc such that $\varphi = \alpha^{-1} \circ \sigma \circ \alpha$.

(a) Show that compositionally similar composition operators are similar.

(b) Let $\varphi(z) = iz$ and $\psi(z) = -iz$. Show that C_φ and C_ψ are similar, but not compositionally similar.

Suggestion: To show that the operators of part (b) are similar, decompose H^2 into the sum of four orthogonal subspaces $\mathcal{H}_j = \{f \in H^2 : \hat{f}(n) = 0 \text{ unless } n \equiv j \pmod 4\}$, where $0 \leq j \leq 3$. Construct the similarity operator S between appropriate subspaces $\mathcal{H}_j$.

10. What happens to the problem above if you don't "multiply by i?" In other words, if $\varphi(z) = z$ and $\psi(z) = -z$, is C_φ similar to C_ψ?

11. Show that similar Hilbert space operators have the same eigenvalues, with the same multiplicities. Then:

(a) Show that different lens maps φ_a $(0 < a < 1)$ induce non-similar composition operators on H^2.

(b) Determine the similarity classes of composition operators induced by linear fractional self-maps of U.

Suggestion: Use the spectral results obtained in the text, and in the problems above.

12. In the course of proving the Eigenfunction Theorem, we proved this result for C_φ^*:

Suppose a bounded linear operator T on a Hilbert space, has an upper triangular matrix with respect to some orthonormal basis $\{e_n\}$ (i.e. $< Te_n, e_m > = 0$ for each $m > n$). Then each diagonal entry of the matrix is an eigenvalue of T.

(a) Show that the diagonal entries of the matrix are the *only* eigenvalues of T.

(b) Show that it is possible for a Hilbert space operator to have a *lower* triangular matrix with respect to the basis $\{z^n\}$, but *no* eigenvalues. (This shows why we had to work with the *adjoint* of C_φ when proving the Eigenfunction Theorem.)

Suggestion: For part (b), consider the operator M_z of multiplication by z on H^2.

6.7   Notes

Königs's Theorem. The local version of Königs's Theorem given above in Exercise 3 plays an important role in complex dynamics; it shows that any holomorphic mapping that has a non-zero derivative λ of modulus < 1 at a fixed point ζ can be "linearized" there in the sense that the original mapping is analytically conjugate in a neighborhood of ζ to its derivative mapping $z \mapsto \lambda z$. This local setting also allows a non-trivial mapping to have derivative of modulus one at the fixed point, something that cannot happen in the "global" context of self-mappings of the unit disc. It is not known exactly which unimodular complex numbers λ have the property that a function f holomorphic in a neighborhood of a fixed point ζ can be linearized there whenever $f'(\zeta) = \lambda$; the issue seems to be one of diophantine approximation. For example, part (b) of Exercise 3 shows that f cannot be linearized if λ is a root of unity. This can also happen— at least for polynomials f—whenever λ belongs to a dense G_δ subset of the circle consisting of points that are sufficiently well-approximated by roots of unity. In the other direction, C.L. Siegel showed in 1942 that there exists a set of unimodular numbers λ having full measure, such that f *is* linearizable at ζ whenever $f'(\zeta) = \lambda$. The point is that these λ's are *badly* approximable by roots of unity. For further references, as well as an exposition of these matters, see Chapter 6 of [Brd '91].

Riesz's Theorem. The results of §6.3 show if a composition operator is compact on H^2, then each non-zero spectral point is an eigenvalue of multiplicity one, and these points form a sequence that converges to zero. Riesz actually showed that something similar holds for *any* compact operator. More precisely, the "real" Riesz Theorem [Rsz '18] asserts, in addition to what we have already stated, that if T is a compact operator on a Hilbert space, then the non-zero eigenvalues (which we observed in Lemma 2 must have finite multiplicity) form a sequence that converges to zero. Our reduced version of Riesz's Theorem showed that if $\lambda \neq 0$ is an eigenvalue of T, then $\bar{\lambda}$ is an eigenvalue of T^*. Riesz showed that in addition, λ and $\bar{\lambda}$ have the same multiplicity. Riesz's work generalized earlier results of Fredholm on integral equations (see for example [RNg '55, §74–§78]), and was in turn generalized to Banach spaces by Schauder [Sdr '30]. For an account of the result in Banach spaces, see [Frd '82, Chapter 4, §5.2], or [Rdn '91, §4.23 – §4.25].

Spectra. The spectrum of a compact composition operator was determined by Caughran and Schwartz [CSc '75], who observed that the same result holds if one only assumes that some power of the operator is compact. This chapter should be viewed as an expanded version of their paper. For the fate of these results in the setting of the unit ball of $\mathbf{C}^n$, see [Mlr '85].

Exercises 5 through 8 show something of the spectral diversity exhibited by non-compact composition operators. Exercises 6 and 8 are results

of Nordgren [Ngn '68], who determined the spectra of composition operators induced by both parabolic (one fixed point) and hyperbolic (two fixed
points) disc automorphisms. These spectra are respectively, the unit circle, and the closure of the annulus in Exercise 8. Exercise 7 comes from
Cowen's paper [Cwn '83] (Cor 6.2). For more on the spectra of composition
operators see [Cwn '83], [Cwn '88], [CwM '92], and [Kwz '75].

Similarity. Exercise 9b shows that the problem of similarity of composition operators is a subtle one. This difficult problem has been taken up
by Campbell-Wright, from whose work Exercise 9b is taken [C-W '91].
In [C-W '93] Campbell-Wright shows that similarity implies compositional
similarity for compact composition operators, as long as the derivatives of
the inducing maps do not vanish at the interior fixed point.

7

Linear Fractional Cyclicity

So far we have concentrated on issues of function theory that were motivated by our studies of boundedness, compactness, and spectra for composition operators. Here we begin exploring the function theoretic-consequences of yet another fundamental notion from operator theory, that of *cyclicity*.

An operator T on a Hilbert space $\mathcal{H}$ is said to be *cyclic* if there is a vector $x \in \mathcal{H}$ such that the orbit $\{T^n x\}_0^\infty$ has dense linear span. When this happens, the vector x is called a *cyclic vector* for T. The notion of cyclicity is intimately connected with the study of invariant subspaces. It is easy to check that the smallest closed T-invariant subspace of $\mathcal{H}$ that contains a given vector is just the closure of the linear span of the orbit of that vector. The best-known problem of Hilbert space theory is that of determining whether or not every bounded operator on Hilbert space has a proper closed (non-trivial) invariant subspace. In the language of cyclicity this problem asks: *Does every Hilbert space operator have a nonzero non-cyclic vector?* This *Invariant Subspace Problem* is, to this point, still unresolved (see the *Notes* at the end of this chapter for more details).

In the next two chapters we are going to show that many composition operators have the strongest possible kind of cyclicity—orbits that are dense *without* requiring help from the linear span. Such operators are called *hypercyclic*, and the vectors that have dense orbits are called *hypercyclic vectors*. There is, of course, no need to limit the notion of hypercyclicity to Hilbert space, or even to vector spaces. The phenomenon occurs frequently in dynamics, where it is called *topological transitivity*. For the record: A continuous self-map T of a topological space is *topologically transitive* if there is a point x in the space such that the orbit $\{T^n x\}_0^\infty$ is dense.

The first linear appearance of hypercyclicity occurred more than seven decades ago in the work of G.D. Birkhoff [Brk '29], who proved that the operator of "translation by one"

$$f(z) \mapsto f(z+1)$$

acts hypercyclically on $H(\mathbf{C})$, the space of entire functions. Here the topology is that of uniform convergence on compact sets, and of course translation by one can be replaced with translation by any fixed non-zero complex number. Seidel and Walsh [SWa '41] later proved a version of Birkhoff's Theorem for $H(U)$, where the relevant operator is *non-Euclidean translation;* in our language, composition with a conformal automorphism of the unit disc having no interior fixed point (and therefore having a unique attractive boundary fixed point that plays the role of ∞).

The goal of this chapter is to investigate the fate of the Seidel-Walsh result in the setting of H^2. We will characterize those linear fractional self-maps of U that induce hypercyclic composition operators. Then in Chapter 8 we will show how further development of the "geometric model" point of view, suggested in Chapter 6 by our study of eigenfunctions, allows results on linear fractional hypercyclicity to be transferred to more general classes of mappings.

7.1 Hypercyclic Fundamentals

The main result of this section is a hypercyclicity criterion which, although stated only for Hilbert space operators, continues to hold (with the same proof) for continuous linear operators on Banach spaces. The proof works as well for complete linear metric spaces (vector spaces that are complete metric spaces in which the operations of vector addition and scalar multiplication are jointly continuous), so the result can be applied in spaces like $H(U)$ and the space of entire functions.

Our insistence that $\mathcal{H}$ be separable, previously used as a psychological crutch, is now absolutely essential; clearly no hypercyclic operator can live on a non-separable space. (Amusingly enough, the same is true for finite-dimensional spaces, see Exercise 4c of this chapter.)

We let $HC(T)$ denote the (possibly empty) set of hypercyclic vectors for T. Upon writing out the definition of $HC(T)$, and exchanging quantifiers for set operations, there emerges the following useful representation for this set. Let S denote any countable dense subset of $\mathcal{H}$. Then

$$HC(T) = \bigcap_{s \in S} \bigcap_k \bigcup_n \{x \in \mathcal{H} : \|T^n x - s\| < 1/k\}. \tag{1}$$

The set in braces is just the inverse image under T^n of the open ball of radius $1/k$, centered at the point s, so it is open because of the continuity of T. The indices k and n range through the positive integers, so formula (1) shows that $HC(T)$ is a countable intersection of open sets, i.e. a G_δ set. In addition, once a point x belongs to $HC(T)$, then so does every point of its orbit $\{T^n x\}$. This proves:

Lemma. *If T is hypercyclic, then $HC(T)$ is a dense G_δ set.*

Since our underlying space is complete, the Baire Category Theorem insures that dense G_δ sets are huge! They are uncountable, and even have uncountable intersection with every open set (since their intersection with every open set is again a dense G_δ). Moreover, every countable intersection of dense G_δ sets is again dense (and certainly G_δ) [Rdn '87, §5.6]. Some consequences for our situation are:

- If an operator is hypercyclic, then "almost every" vector is a hypercyclic vector.

- Every countable collection of hypercyclic operators has a *common* hypercyclic vector.

All the results of this chapter will follow in one way or another from the next theorem, which establishes a sufficient condition for hyperclicity. Here the statement "$T^n \to 0$ on a set X" means that $\|T^n x\| \to 0$ for every vector $x \in X$.

The Hypercyclicity Criterion. *Suppose there is a dense subset X of $\mathcal{H}$ on which $T^n \to 0$, and another dense set Y on which is defined a (possibly discontinuous) map $S : Y \to Y$ such that*

(a) *TS is the identity on Y, and*

(b) *$S^n \to 0$ on Y.*

Then T is hypercyclic.

Despite its complicated appearance, this result is widely applicable, and often surprisingly easy to use. To make this point we defer the proof in order to present two important examples. We begin with what is historically the first example of a hypercyclic *Hilbert space* operator. For this, recall the *backward shift* operator B that played an important role in our proof of Littlewood's Subordination Principle (§1.3). Since B is a contraction on H^2, it is certainly not hypercyclic, but the next result shows that it becomes so with just a little extra help.

Rolewicz's Theorem. *The operator λB is hypercyclic on H^2 for every complex number λ of modulus > 1.*

Proof. Let $T = \lambda B$. Our goal is to find the dense sets X and Y, and the mapping S requested by the Hypercyclicity Criterion. Let X be the collection of polynomials; clearly a dense subset of H^2. Now if $f \in X$ has degree N, then for each $n > N$ we have $B^n f = 0$, and therefore $T^n f = 0$. Thus (trivially) $T^n \to 0$ on X.

To finish the proof, let Y be H^2 itself, and consider the operator of multiplication by z:

$$M_z f(z) = z f(z) \qquad (f \in H^2).$$

Since M_z shifts the coefficients of f one unit to the right, the operator BM_z is the identity on H^2. Now define define $S = \lambda^{-1} M_z$. Then TS is the identity on H^2, and for each $f \in H^2$ and each positive integer n we have

$$\|S^n f\| = |\lambda|^{-n}\|zf\| = |\lambda|^{-n}\|f\| \to 0 \quad (n \to \infty),$$

where at the last step we used the fact that M_z is an isometry of H^2, and (finally) the fact that $|\lambda| > 1$. The result now follows from the Hypercyclicity Criterion. $\quad\square$

Our second example foreshadows what is to come; it is the H^2 version of the Seidel-Walsh Theorem.

Theorem. *Suppose φ is a conformal automorphism of the unit disc with no fixed point in the interior. Then C_φ is hypercyclic on H^2.*

Proof. As discussed in §0.4, the automorphism φ, being non-elliptic, has a unique attractive fixed point $\alpha \in \partial U$. If there is another fixed point β, then this too must lie on the unit circle since it is the attractive fixed point for the inverse of φ, which is again an automorphism of U. Suppose first that we are dealing with the case of two fixed points. Just as in the proof of Rolewicz's Theorem, our goal is to produce the cast of characters required for the hypothesis of the Hypercyclicity Criterion.

Let X denote the set of functions that are continuous on the closed unit disc, analytic on the interior, and which vanish at α. We claim that $C_\varphi^n \to 0$ on X. For this, note that for every $\zeta \in \partial U \backslash \{\beta\}$ we have $\varphi_n(\zeta) \to \alpha$, hence if $f \in X$ then $f(\varphi_n(\zeta)) \to f(\alpha) = 0$. Upon applying the elementary case of the boundary integral representation of the H^2 norm ((1) of §2.3), the Lebesgue Bounded Convergence Theorem, yields the desired result:

$$\|C_\varphi^n f\|^2 = \frac{1}{2\pi} \int_{-\pi}^{\pi} |f(\varphi_n(e^{i\theta}))|^2 d\theta \to 0 \qquad (n \to \infty).$$

There are several ways to see that X is dense in H^2. Here is one based on elementary Hilbert space theory. Suppose $f \in H^2$ is orthogonal to X. Then for every non-negative integer n, the polynomial $z^{n+1} - \alpha z^n$ belongs to X, so it is orthogonal to f:

$$0 = < f, z^{n+1} - \alpha z^n > = \hat{f}(n+1) - \bar{\alpha}\hat{f}(n).$$

It follows upon iterating this identity that $\hat{f}(n) = \bar{\alpha}^n \hat{f}(0)$ for each n. Since α has modulus one, all the Taylor coefficients of f have the same modulus, and since $f \in H^2$, these coefficients must all be zero. Thus the only H^2 function orthogonal to X is the zero function. Since X is a linear subspace of H^2, it must therefore be dense.

We note for further reference that the only property required here of α is that it lie outside of U; the argument actually shows:

> If $\alpha \notin U$ then the set of polynomials that vanish at α is dense in H^2.

To finish the proof let $S = C_\varphi^{-1} = C_{\varphi^{-1}}$. As noted above, φ^{-1} is also an automorphism of the disc, with attracting fixed point β (the repulsive fixed point of φ). So if we take Y to be the set of continuous functions on the disc that are holomorphic in the interior and vanish at β, then S maps Y into itself, and the previous arguments apply to show that Y is dense and $S^n \to 0$ on Y. The hypotheses of the Hypercyclicity Criterion are therefore satisfied, so C_φ is hypercyclic on H^2.

The case where φ has just one fixed point is even easier; take X as before, and set $Y = X$. We leave the details to the reader.   □

A comparison principle. The Seidel-Walsh Theorem is actually a special case of the last result. To see this we need only note that H^2 is dense in $H(U)$ (the polynomials are dense in both), so because convergence in H^2 implies convergence in $H(U)$ (§1.1), any hypercyclic vector for C_φ acting on H^2 is also hypercyclic for C_φ acting on $H(U)$. Clearly the argument given here works in far more generality, and yields:

The Hypercyclic Comparison Principle. *Suppose E is a linear metric space, and F a dense subspace that is itself a linear metric space with a stronger topology. Suppose T is a linear transformation on E that also maps the smaller space F into itself, and is continuous in the topology of each space. If T is hypercyclic on F, then it is also hypercyclic on E, and has an E-hypercyclic vector that belongs to F.*

Results like this show why it may be of interest to generalize known hypercyclicity theorems to smaller spaces. Success produces "better behaved" hypercyclic vectors for the original theorem. For example, the proof of the automorphic hypercyclicity theorem works, with suitable modifications, for any space H^p with $p < \infty$, and this establishes that in the original

Seidel-Walsh Theorem there is a hypercyclic vector in each H^p. With a little more care it can even be shown that there is a hypercylic vector that belongs to $\cap_{p<\infty} H^p$. On the other hand, it is easy to see that no hypercyclic vector can belong to H^∞. For an application of this idea that is related to Birkhoff's Theorem, see [ChS '91].

Proof of the Hypercyclicity Criterion. We have to show that, under the hypotheses of the theorem, the set of hypercyclic vectors is non-empty. Now $HC(T)$ is exhibited in (1) above as a countable intersection of open sets of the form

$$G(y,\varepsilon) = \{x \in \mathcal{H} : \|T^n x - y\| < \varepsilon \text{ for some } n\},$$

where $y \in \mathcal{H}$ and $\varepsilon > 0$. By the completeness of $\mathcal{H}$, Baire's Theorem [Rdn '87, §5.6] will insure that this intersection is non-empty, once we prove that each of the above sets is dense in $\mathcal{H}$.

Given y and ε, choose any $x_0 \in \mathcal{H}$ and $\delta > 0$. Our task is to show that the ball of radius δ about x_0 contains a point of $G(y,\varepsilon)$.

Since $T^n \to 0$ on a dense subset of $\mathcal{H}$ there is a point $x_1 \in \mathcal{H}$ such that $\|x_1 - x_0\| < \delta/2$, and $\|T^n x_1\| < \varepsilon/2$ for all sufficiently large n. Since also $S^n \to 0$ on a dense subset of $\mathcal{H}$, there is a point $y_1 \in \mathcal{H}$ such that $\|y_1 - y\| < \varepsilon/2$ and $\|S^n y_1\| < \delta/2$ for all sufficiently large n.

We claim that whenever n is "sufficiently large," in the sense of the last paragraph, the point

$$x = x_1 + S^n y_1$$

is the one we seek. Indeed,

$$\|x - x_0\| \le \|x_1 - x_0\| + \|S^n y_1\| < \delta/2 + \delta/2 = \delta,$$

so x has the first of the desired properties. To see that it lies in $G(y,\varepsilon)$, note that since TS is the identity map on Y, the same is true of $T^n S^n$ (even though we are *not* assuming that T and S commute). The linearity of T (used here for the first time) now shows that

$$T^n x = T^n x_1 + T^n S^n y_1 = T^n x_1 + y_1$$

from which follows

$$\|T^n x - y\| \le \|T^n x_1\| + \|y_1 - y\| < \varepsilon/2 + \varepsilon/2 = \varepsilon.$$

Thus x belongs to $G(y,\varepsilon)$, and lies within δ of x_0, as desired.   □

7.2   Linear Fractional Hypercyclicity

Having proved hypercyclicity for the composition operators induced by automorphisms of U with no interior fixed point, we turn next to the class of operators induced by all linear fractional self-maps of U; the so-called *linear fractional* composition operators. Our previous experience suggests the inducing map's behavior at its fixed points can be expected to play a significant role here. With this in mind we dispose, once and for all, of maps with interior fixed points.

Proposition 1. *If φ is a holomorphic self-map of U that fixes a point of U, then C_φ is not hypercyclic on H^2.*

Proof. Suppose $p \in U$ is a fixed point for φ. Fix $f \in H^2$, to be regarded as a "hypercyclic vector candidate." If $g \in H^2$ belongs to the closure of the orbit of f, then for some subsequence $n_k \nearrow \infty$ we have $f \circ \varphi_{n_k} \to g$ in the norm of H^2, and therefore by the work of §1.1, pointwise on U. Because $\varphi(p) = p$ it follows that

$$g(p) = \lim_k f(\varphi_{n_k}(p)) = f(p).$$

We conclude that the closure of the orbit of f excludes every H^2 function that does not agree with f at the fixed point p, so no orbit can be dense in H^2. $\square$

For this proof we could also have used Littlewood's Subordination Principle, which asserts that if the fixed point is the origin, then C_φ is a contraction, so clearly not hypercyclic (the orbit of each vector f has to stay in the closed ball of radius $\|f\|$ about the origin). If $p \neq 0$ then we have, as in Chapter 1, $\varphi = \alpha_p \circ \psi \circ \alpha_p$, where $\psi(0) = 0$, so by Exercise 9 of §6.6, C_φ is similar to a contraction. It is an easy exercise to check that similarity preserves hypercyclicity, hence we conclude once more that C_φ is not hypercyclic.

Recall that we saw at the end of Chapter 5 that if a composition operator is compact, then its inducing map has an interior fixed point. This result, along with the one above proves:

Corollary. *No compact composition operator is hypercyclic.*

In fact much more is true:

No compact operator on Hilbert space is hypercyclic,

and this result is even true for Banach spaces (see [Kit '82] for the details).

From the work of §0.4, the only members of $LFT(U)$ that have even a chance to induce hypercyclic composition operators are the parabolic and hyperbolic ones with attractive fixed point on the boundary. The main result of this chapter is that "most" of these mappings succeed, but the others fail miserably.

The Linear Fractional Hypercyclicity Theorem. *Suppose that $\varphi \in LFT(U)$ has no fixed point in U. Then:*

(a) *C_φ is hypercyclic on H^2 unless φ is a parabolic non-automorphism.*

(b) *If φ is a parabolic non-automorphism, then C_φ fails to be hypercyclic in a very strong sense: Only constant functions can be limit points of C_φ orbits.*

Proof of (a). We have already proved the result for automorphisms, so it remains to do it for hyperbolic non-automorphisms. Let φ be such a map, and suppose α and β are its fixed points, with α the attractive one. As before we seek to find the dense sets X and Y, and the map S that will satisfy the hypotheses of the Hypercyclicity Criterion.

The space X is exactly the one that worked in the automorphic hypercyclicity result of the previous section, and it works again with no change in the argument. It is the space Y that requires some care.

Suppose first that the repulsive fixed point β lies on the line through the origin and α, but is on the other side of the origin from α. Let Δ be the disc whose boundary is the circle perpendicular to this line that passes through α and β (see Figure 7.1 below).

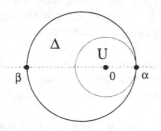

FIGURE 7.1. *The discs U and Δ*

Since φ fixes α and β and preserves angles, it maps the boundary of Δ onto itself, and therefore takes Δ onto either itself or the exterior of ∂D. But φ takes U into itself, so the latter possibility is ruled out. Therefore φ is a conformal automorphism of Δ.

Let Y be the collection of functions that are continuous on $\overline{\Delta}$, analytic on Δ, and which vanish at β. As we noted in the last section, the fact that β lies outside U insures that the polynomials that vanish at β form a dense subset of H^2, thus Y is dense. Define the map $S : Y \to Y$ by

$$Sf(z) = f(\varphi^{-1}(z)) \qquad (z \in U).$$

The fact that $\varphi^{-1}(z)$ is not always in U is of no importance here, nor is the fact that S is neither defined nor bounded on H^2. What is important is that $\varphi^{-1}(\Delta) \subset \Delta$, that S is defined on Y, and that SC_φ is the identity on Y, all of which is obvious. In addition, the fact that $\varphi^{-n}(z) \to \beta$ for each $\zeta \in \partial U$ yields, precisely as in the last section, that $S^n \to 0$ on Y. Thus the hypotheses of the Hypercyclicity Criterion are again satisfied, so C_φ is hypercyclic on H^2.

If the repulsive fixed point β is not in the required position (it could even be at ∞ for example), then there is a conformal automorphism γ of U that fixes α and takes β to the desired position (see Exercise 10 of §7.4 below). Then $\varphi = \gamma \circ \psi \circ \gamma^{-1}$, where ψ is a linear fractional self-map of U that has its fixed points arranged properly, so C_ψ is hypercyclic. Now the argument that provided the alternate proof of Proposition 1 above strikes again, and shows that C_φ is similar to C_ψ, and therefore hypercyclic.   □

Remark. Once it has been observed that φ is an automorphism of the larger disc Δ, a more elegant line of argument suggests itself. Define the space $H^2(\Delta)$ in some obvious way, show that it is a dense subspace of H^2 and has a stronger topology. Then use the Automorphism Theorem to conclude that C_φ is hypercyclic on $H^2(\Delta)$, and transfer this hypercyclicity to H^2 itself by means of the Comparison Theorem of the previous section. All of this works; the idea is exploited in far more generality in the next chapter.

Remark. It is interesting to investigate why the proof given above for hyperbolic non-automorphisms does not work for parabolic ones. The point is that in the hyperbolic case the big disc Δ is precisely the union of the successive inverse images of U under φ:

$$\Delta = \bigcup_n \varphi^{-n}(U).$$

If, on the other hand, φ is a *parabolic* non-automorphism, then this union turns out to be $\widehat{C} \backslash \{\alpha\}$ (as is easily seen by representing the map as a translation of the right half-plane strictly into itself), hence the set Y defined in the proof above contains only the zero function.

Proof of (b). Now we assume that $\varphi \in LFT(U)$ is a parabolic non-automorphism, so it has only one fixed point in $\widehat{C}$, and this lies on ∂U. Without loss of generality we may take this fixed point to be $+1$ (otherwise conjugate φ by an appropriate rotation to produce a similar composition operator induced by a parabolic automorphism with fixed point at $+1$).

In §0.4 we obtained the following formula for φ (valid for both automorphisms and non-automorphisms):

$$\varphi(z) = 1 + \frac{2(z-1)}{2 - a(z-1)} \, , \tag{2}$$

where $a = \varphi''(1)$. In this case, a has strictly positive real part, because φ is not an automorphism.

To obtain the corresponding formula for the iterate φ_n, it is best to recall the derivation of (2). Under the transformation $w = (1 + z)/(1 - z)$, the original map φ became the translation $\Phi(w) = w + a$ of the right half-plane, which showed right away that $\mathrm{Re}\, a$ had to be non-negative in general, and strictly positive for non-automorphisms. Since the half-plane model that corresponds to φ_n is just translation by na, the formula for φ_n is given by (2) with a replaced by na.

Our proof will depend on knowing how quickly the φ-orbits of points in U get close to each other, and to the attractive fixed point $+1$. With this in mind we use (2) to compute

$$\varphi(z) - \varphi(0) = \frac{4z}{(2+a)(2+a-az)}. \tag{3}$$

Upon replacing a by na in (2) and (3), and letting $n \to \infty$ we obtain precise asymptotic information on the quantities of interest:

$$\lim_{n \to \infty} n[1 - \varphi_n(z)] = \frac{2}{a}, \tag{4}$$

$$\lim_{n \to \infty} n^2[\varphi_n(z) - \varphi_n(0)] = \frac{4z}{a^2(1 - z)}. \tag{5}$$

Now fix $f \in H^2$. Our goal is to show that if the orbit of f under C_φ clusters at some $g \in H^2$, then g must be a constant function. For this we need a growth estimate on *differences* of functional values that is analogous the one obtained in §1.1 for the values themselves. We begin with the derivative. For $z \in U$,

$$\begin{aligned}
|f'(z)|^2 &= \left| \sum_{n=1}^{\infty} n\hat{f}(n)z^{n-1} \right|^2 \\
&\leq \left(\sum_{n=1}^{\infty} |\hat{f}(n)|^2 \right) \left(\sum_{n=1}^{\infty} n^2 |z|^{2(n-1)} \right) \\
&\leq \|f\|^2 \frac{2}{(1 - |z|^2)^3}.
\end{aligned}$$

Upon taking square roots on both sides of the last inequality, we get this growth estimate on the derivative of f:

$$|f'(z)| \leq \sqrt{2} \frac{\|f\|}{(1 - |z|^2)^{3/2}} \qquad (z \in U).$$

To get the desired estimate on differences, suppose $z, w \in U$ and $|z| \leq |w|$. To estimate $f(z) - f(w)$ we integrate f' over the line segment joining z and w, and use the inequality above:

$$|f(z) - f(w)| \leq \int_z^w |f'(\zeta)| |d\zeta|$$

$$\leq \sqrt{2} \|f\| \int_z^w \frac{|d\zeta|}{(1 - |\zeta|^2)^{3/2}}$$

$$\leq \sqrt{2} \|f\| \frac{|w - z|}{(1 - |w|^2)^{3/2}}.$$

Thus for each pair of points $z, w \in U$,

$$|f(z) - f(w)| \leq \sqrt{2} \|f\| \frac{|w - z|}{(\min\{1 - |w|, 1 - |z|\})^{3/2}}. \qquad (6)$$

The final piece of the puzzle involves the geometry of φ-orbits. The half-plane representation of φ is $\Phi(w) = w + a$, so that of φ_n is $\Phi_n(w) = w + na$. The fact that $\operatorname{Re} a > 0$ (now used for the first and only time in the proof!) forces the points $\Phi_n(w)$ to approach ∞ along a line in the right half-plane that is not parallel to the imaginary axis. Back in the disc this forces every orbit $\{\varphi_n(z)\}$ to approach the attractive fixed point $+1$ non-tangentially.

Now fix $z \in U$. The non-tangential convergence of orbits provides a constant $c > 0$ such that for each n,

$$1 - |\varphi_n(z)| \geq c|1 - \varphi_n(z)| \quad \text{and} \quad 1 - |\varphi_n(0)| \geq c|1 - \varphi_n(0)|. \qquad (7)$$

In (6) above, substitute $\varphi_n(z)$ for z, and $\varphi_n(0)$ for w, and use successively (7), and (4) along with (5); the result is

$$|f(\varphi_n(z)) - f(\varphi_n(0))| \leq \text{const.} \frac{|\varphi_n(z) - \varphi_n(0)|}{(\min\{1 - |\varphi_n(z)|, 1 - |\varphi_n(0)|\})^{3/2}}$$

$$\leq \text{const.} \frac{n^{-2}}{n^{-3/2}}$$

$$= \frac{\text{const.}}{\sqrt{n}},$$

where the constant in each line depends on f, z, and φ, but not on n. Thus

$$\lim_n [f(\varphi_n(z)) - f(\varphi_n(0))] = 0 \qquad (z \in U). \qquad (8)$$

To finish the argument, suppose $g \in H^2$ is a cluster point of the C_φ -orbit of f. Then for some sequence $n_k \nearrow \infty$ we have $f \circ \varphi_{n_k} \to g$ in the norm of H^2, and therefore pointwise on U. By (8) this implies

$$g(z) - g(0) = \lim_k [f(\varphi_{n_k}(z)) - f(\varphi_{n_k}(0))] = 0,$$

hence $g \equiv g(0)$. Thus only constant functions can be limit points of the C_φ-orbit of an H^2 function.   □

7.3   Linear Fractional Cyclicity

So far we have seen that if $\varphi \in LFT(U)$ fixes no interior point, then the composition operator it induces is hypercyclic on H^2, except when φ is a parabolic non-automorphism. There is still the possibility that even in this exceptional case C_φ may be *cyclic*. In other words, even though there are no dense orbits, it is still possible that the *linear span* of an orbit is dense. In this section we show that this is exactly what happens.

The usual strategy for proving that a vector is cyclic for a Hilbert space operator is to show that only the zero vector can be orthogonal to the orbit of the given vector (recall that once a vector is orthogonal to a set, then it is also orthogonal to the span of that set). A similar idea was employed in the last two sections to prove density for the subspaces X and Y required by the Hypercyclicity Theorem. Behind this strategy there is usually a classical uniqueness theorem; in our previous application it was the fact that analytic functions are determined by their Taylor coefficients. In this one it will be the fact that H^2 *functions cannot have zeros that are too close together.*

The required estimate on the placement of zeros depends on *Jensen's Formula*, which will also play a crucial role in Chapter 10. To discuss this topic we employ the following terminology. The *zero-sequence* of a function f holomorphic in U is the collection of its zeros, listed in order of increasing moduli, with each zero written down as many times as its multiplicity. For $0 \le r < 1$ we write $n(r)$ for the number terms of the zero-sequence of f with modulus $\le r$.

For example, if f has a zero of order n at the point $1 - n^{-1}$ $(n \ge 2)$, then the zero-sequence of f is

$$\{\frac{1}{2}, \frac{1}{2}, \frac{2}{3}, \frac{2}{3}, \frac{2}{3}, \frac{3}{4}, \frac{3}{4}, \frac{3}{4}, \frac{3}{4}, \ldots\},$$

and $n(2/3) = 5$. The results to follow will show that no function of class H^2 has this as its zero-sequence.

Jensen's Formula. *Suppose $f \in H(U)$ and $f(0) \ne 0$. Let $\{a_n\}_1^\infty$ be the zero-sequence of f. Then for each $0 \le r < 1$,*

$$\sum_{n=1}^{n(r)} \log \frac{r}{|a_n|} = \frac{1}{2\pi} \int_{-\pi}^{\pi} \log |f(re^{i\theta})| \, d\theta - \log |f(0)|. \tag{9}$$

Proof. If $0 \leq r < |a_1|$ the sum is zero, and the result is just the mean value property of the function $\log |f|$, which is harmonic in a neighborhood of the closed disc $r\overline{U}$.

Suppose $r \geq |a_1|$. If f has no zero of modulus r, then for $0 \leq n \leq n(r)$ we have $a_n/r \in U$. Let α_n denote the special automorphism that interchanges the point a_n/r with the origin:

$$\alpha_n(z) = \alpha_{a_n/r}(z) = \frac{a_n - rz}{r - \overline{a}_n z},$$

and set

$$h(z) = \prod_{n=1}^{n(r)} \alpha_n(z).$$

Then, in a neighborhood of the closed unit disc, $h(z)$ is holomorphic, and has the same zeros as $f(rz)$ (with the same multiplicities). Thus $f(rz)/h(z)$ is holomorphic and non-vanishing in a neighborhood of $\overline{U}$, so $\log |f(rz)/h(z)|$ is harmonic there. By the mean value property,

$$\log \left| \frac{f(0)}{h(0)} \right| = \frac{1}{2\pi} \int_{-\pi}^{\pi} \log \left| \frac{f(re^{i\theta})}{h(e^{i\theta})} \right| \, d\theta = \frac{1}{2\pi} \int_{-\pi}^{\pi} \log |f(re^{i\theta})| \, d\theta$$

where the last equality holds because $|h| \equiv 1$ on ∂U. Since

$$h(0) = \prod_{1}^{n(r)} \frac{r}{a_n},$$

this is the desired identity.

Suppose f *does* have zeros of modulus r, say

$$|a_{k+1}| = |a_{k+2}| = \cdots = |a_{n(r)}| = r,$$

and all the other zeros in rU (if there are any) have modulus $< r$. Now apply the argument above, but this time with

$$h(z) = \prod_{n=1}^{k} \alpha_n(z) \prod_{n=k+1}^{n(r)} \left(1 - \frac{rz}{a_n}\right).$$

Reasoning as in the paragraph above, we obtain:

$$\log \left| \frac{f(0)}{h(0)} \right| = \frac{1}{2\pi} \int_{-\pi}^{\pi} \log \left| \frac{f(re^{i\theta})}{h(e^{i\theta})} \right| \, d\theta$$

$$= \frac{1}{2\pi} \int_{-\pi}^{\pi} \log |f(re^{i\theta})| \, d\theta - \sum_{n=k+1}^{n(r)} \frac{1}{2\pi} \int_{-\pi}^{\pi} \log |1 - e^{(i\theta - \theta_n)}| \, d\theta,$$

where in the sum on the right, θ_n is the argument of a_n. We claim that each term of this sum is zero, i.e., that

$$\int_{-\pi}^{\pi} \log|1 - e^{i\theta}|\, d\theta = 0.$$

Most textbooks use a residue calculation to obtain this (cf [Rdn '87, §15.17], but a completely elementary argument suffices [Yng '86]. Using symmetry and the change of variable $t = \theta/2$ the integral is equal to

$$4\int_0^{\pi/2} \log(2\sin t)\, dt\ ,$$

so it is enough to show that

$$I \stackrel{\text{def}}{=} \int_0^{\pi/2} \log(\sin t)\, dt = -\frac{\pi}{2}\log 2\ .$$

Now the double-angle formula yields

$$I \ = \ \int_0^{\pi/2} \log(2\sin\frac{t}{2}\cos\frac{t}{2})\, dt$$

$$= \ \frac{\pi}{2}\log 2 + \int_0^{\pi/2}\log\sin\frac{t}{2}\, dt + \int_0^{\pi/2}\log\cos\frac{t}{2}\, dt\ .$$

Upon substituting $\theta = t/2$ in the first integral above, and $\theta = (\pi - t)/2$ in the second, we obtain

$$I = \frac{\pi}{2}\log 2 + 2\int_0^{\pi/2}\log\sin\theta\, d\theta = \frac{\pi}{2}\log 2 + 2I,$$

hence $I = -\frac{\pi}{2}\log 2$, as desired.   □

Zero-Sequence Theorem. *Suppose $\{a_n\}_1^\infty$ is the zero-sequence of a function $f \in H^2$ that is not identically zero. Then*

$$\sum_{n=1}^{\infty}(1 - |a_n|) < \infty.$$

Proof. We may assume without loss of generality that f has infinitely many zeros, and that $f(0) = 1$. Fix a positive integer N. Then for any $0 < r < 1$ for which $n(r) > N$ (this happens for all r sufficiently close to 1) we use respectively Jensen's Formula and the Arithmetic-Geometric Mean Inequality [Rdn '87, §3.3, inequality (7)] to obtain:

$$\sum_{n=1}^{N} \log \frac{r}{|a_n|} \;\leq\; \sum_{n=1}^{n(r)} \log \frac{r}{|a_n|}$$

$$= \frac{1}{2\pi} \int_{-\pi}^{\pi} \log |f(re^{i\theta})| \, d\theta$$

$$\leq \log \left\{ \frac{1}{2\pi} \int_{-\pi}^{\pi} |f(re^{i\theta})|^2 d\theta \right\}^{1/2}$$

$$\leq \log \|f\|,$$

where the norm in the last line is that of H^2.

Now let $r \to 1-$ on the right-hand side of the string of inequalities above, and then let $N \to \infty$. The result is:

$$\sum_{n=1}^{\infty} \log \frac{1}{|a_n|} \leq \log \|f\|,$$

which, because $1 - x \leq -\log x$ for $x > 0$, yields the desired result. □

Theorem (Linear Fractional Cyclicity). *If $\varphi \in LFT(U)$ fixes no point of U, then C_φ is cyclic on H^2.*

Proof. As mentioned above, we need only consider $\varphi \in LFT(U)$ a parabolic non-automorphism. Just as in the last section, we may, without loss of generality, suppose that the fixed point of φ is at $+1$. We are going to show that the *identity function* $u(z) = z$ is a cyclic vector for C_φ.

We start off with the representation (2) for φ that was employed in the last section, and rewrite it as

$$\varphi(z) = \frac{a + (2 - a)z}{(2 + a) - az} .$$

A standard way of decomposing linear fractional transformations involves subtracting off the value at ∞ and rearranging what remains into something that looks like a constant multiple of a reproducing kernel. When applied to the formula above, this procedure yields:

$$\varphi(z) = q + ck_p \tag{10}$$

where

$$q = \varphi(\infty) = \frac{a - 2}{a}, \quad c = \frac{4}{a(2 + a)}, \quad p = \frac{\bar{a}}{2 + \bar{a}} ,$$

and $k_p(z) = (1 - \bar{p}z)^{-1}$. Because $\operatorname{Re} a > 0$, none of the denominators above vanish, $c \neq 0$, and $p \in U$. Thus k_p is the reproducing kernel for p (see the end of §3.4), and so

$$< f, k_p >= f(p) \qquad (f \in H^2).$$

Now suppose $f \in H^2$ is orthogonal to the C_φ-orbit of the identity function u, that is,

$$< f, \varphi_n >= 0 \qquad (n = 0, 1, 2, \ldots).$$

We claim that f is therefore orthogonal to each constant function (less pretentiously: $f(0) = 0$). Indeed, on ∂U the iterates $\{\varphi_n\}$ are uniformly bounded and converge pointwise to $+1$, so by the boundary integral representation of the H^2 norm (actually, of the *inner product*), and the Lebesgue Dominated Convergence Theorem,

$$\begin{aligned}
0 &= \lim_n < f, \varphi_n > \\[2ex]
&= \lim_n \frac{1}{2\pi} \int_{-\pi}^{\pi} f(e^{i\theta}) \overline{\varphi_n(e^{i\theta})} d\theta \\[2ex]
&= \frac{1}{2\pi} \int_{-\pi}^{\pi} f(e^{i\theta}) d\theta \\[2ex]
&= < f, 1 >,
\end{aligned}$$

as promised. This result, the reproducing property of k_p, and the representation (10) of φ now combine to show that

$$0 =< f, \varphi >=< f, q + ck_p >= \bar{c} < f, k_p >= \bar{c} f(p),$$

so, because $c \neq 0$, we have $f(p) = 0$.

This is the key to the argument. Since the formula for the iterate φ_n is obtained from that of φ by replacing the translation vector a by na, we see that φ_n has representation

$$\varphi_n = q_n + c_n k_{p_n},$$

where q_n and c_n are complex constants, and

$$p_n = \frac{n\bar{a}}{2 + n\bar{a}} \in U.$$

Thus the work of the last paragraph also shows that $f(p_n) = 0$ for all non-negative integers n, so to conclude the proof we need only show that among H^2 functions, only the zero function can vanish identically on the sequence $\{p_n\}$. For this we appeal one last time to the strict positivity of Re a to obtain the estimate

$$1 - |p_n|^2 = \frac{4(1 + \text{Re}\, na)}{4(1 + \text{Re}\, na) + n^2|a|^2} \geq \frac{\text{const.}}{n}.$$

Thus $\sum_n (1 - |p_n|) = \infty$, so the desired result follows from the Zero-Sequence Theorem.   □

This completes the classification of cyclic behavior for composition operators induced by linear fractional mappings with no interior fixed point. We summarize the situation in the Table below, where, to make statements about derivatives as simple as possible, the boundary fixed point is assumed to be at $+1$.

Type of φ	Hyperbolic	Parabolic automorphism	Parabolic non-auto.
Cyclicity of C_φ	Hypercyclic	Hypercyclic	Cyclic (not hypercyclic.)
Derivatives at attractive F.P. ($\alpha = +1$)	$0 < \varphi'(1) < 1$	$\varphi'(1) = 1$ $\text{Re}\,\varphi''(1) = 0$	$\varphi'(1) = 1$ $\text{Re}\,\varphi''(1) > 0$
Right half-plane Model	$\Phi(w) = \lambda w + b$ $\lambda = \varphi'(1)$	$\Phi(w) = w + b$ $b = \varphi''(1)$	

Cyclicity of C_φ for $\varphi \in LFT(U)$, no fixed point in U.

The Seidel-Walsh Theorem, revisited. Since the proof of the last theorem depended critically on the fact that the derivatives of H^2 functions have suitably restricted growth, it leaves open the possibility that composition operators induced by parabolic non-automorphisms might still be hypercyclic on the full space of holomorphic functions on U. Restating the question:

Does the Seidel-Walsh Theorem hold for every $\varphi \in LFT(U)$ with no interior fixed point?

The answer is "Yes," and this can be seen as follows. Suppose φ is a parabolic non-automorphism whose fixed point is $+1$. We have seen that upon taking the unit disc to the right half-plane by the transformation $w = (1 + z)/(1 - z)$, the map φ goes over to the translation mapping $\Phi(w) = w + a$, where $\text{Re}\, a > 0$, so C_φ corresponds to the operator of translation by a,

$$T_a f(w) = f(w + a) \qquad (w \in \Pi),$$

acting on $H(\Pi)$. Thus hypercyclicity of C_φ on $H(U)$ is equivalent to the right half-plane version of Birkhoff's Theorem, which we recall asserts the hypercyclicity of T_a on $H(\mathbf{C})$. By Runge's Theorem, $H(\mathbf{C})$ is dense in $H(\Pi)$, and clearly the smaller space has the stronger topology. Therefore the Comparison Principle of §7.1 insures that T_a is hypercyclic on $H(\Pi)$. $\square$

We will not prove Birkhoff's Theorem here. For a proof based on the Hypercyclicity Criterion of §7.1, see [GrS '87] or [GyS '91, §5].

The interior fixed point case. If $\varphi \in LFT(U)$ has a fixed point in U then we saw in §7.2 (Proposition 1) that C_φ cannot be be hypercyclic. But it still might be cyclic. For example, Exercise 11 below shows that this happens whenever φ is elliptic and the argument of its multiplier is an irrational multiple of π, or whenever φ is loxodromic. The rest of the picture was sketched out in [BoS '90], and discussed more fully in [BoS '93]. It turns out that for φ non-elliptic, but with a fixed point in U, the operator C_φ will be cyclic whenever the other fixed point is outside $\overline{U}$, but will be *non-cyclic* (in fact, very strongly so), if this fixed point is on ∂U (see [BoS '90] and [BoS '93] for the details).

7.4 Exercises

1. Show that no multiple of the forward shift M_z is hypercyclic on H^2.

2. Show that the backward shift B is hypercyclic on the Bergman space A^2 defined in Exercise 4 of Chapter 1 (note that no scalar multiple is needed here!)

3. *MacLane's Theorem.* Show that the operator of differentiation is hypercyclic on the space $H(\mathbf{C})$ of entire functions, in the topology of uniform convergence on compact subsets of the plane.

   Suggestion: In the Hypercyclicity Criterion of §7.1, take $X = Y =$ the space of (holomorphic) polynomials.

In the next six problems, you are invited to prove the assertions made about T, a bounded operator on a Hilbert space $\mathcal{H}$.

4. (a) If T^* has an eigenvalue, then T is not hypercyclic.

   (b) No operator on a finite dimensional Hilbert space is hypercyclic.

   (c) If F is a finite rank operator on $\mathcal{H}$, then $I + F$ is not hypercyclic.

5. If T^* has an eigenvalue of multiplicity at least *two*, the T is not cyclic.

   Suggestion: The closure of the range of a cyclic operator can have codimension at most one.

6. If T is hypercyclic, then every vector in $\mathcal{H}$ is the sum of two hypercyclic vectors.

   Suggestion: The hypercyclic vectors form a dense G_δ set.

7. *Sensitive dependence on initial conditions.* If T is hypercyclic then for every $x \in \mathcal{H}$ and every $\varepsilon > 0$, the ball of radius ε centered at x contains an uncountable dense set of vectors y such that

$$\limsup_{n \to \infty} \|T^n x - T^n y\| = \infty.$$

8. *Strong hypercyclicity.* If T satisfies the Hypercyclicity Criterion, then for each subsequence $\{n_k\}$ of positive integers there is a dense G_δ set of vectors x such that the set $\{T^{n_k} f\}_{k=0}^\infty$ is dense.

9. If T_1 and T_2 satisfy the Hypercyclicity Criterion on Hilbert spaces $\mathcal{H}_1$ and $\mathcal{H}_2$ respectively, then the direct sum operator $T_1 \oplus T_2$ is hypercyclic on $\mathcal{H}_1 \oplus \mathcal{H}_2$.

   Remark: In general the direct sum of two hypercyclic operators need not even be cyclic [Sls '91]. It is not known if the direct sum of a hypercyclic operator with itself has to be hypercyclic.

10. In the proof of the Linear Fractional Hypercyclicity Theorem we needed to be able to use disc automorphisms to move fixed points around. Suppose $\alpha \in \partial U$ and $\beta \in \widehat{\mathbf{C}} \backslash \overline{U}$. Show that there is a conformal automorphism of U that takes β onto the part of the line through α and 0 that lies on the opposite side of the origin from β.

    Suggestion: Without loss of generality you can take $\alpha = 1$. Use the map $w = (1 + z)/(1 - z)$ to replace the unit disc by the right half-plane, and α by ∞. An appropriate affine map does the job in the half-plane.

11. Suppose $\varphi \in LFT(U)$ has a fixed point α in U.

    (a) Suppose φ is elliptic ($|\varphi'(\alpha)| = 1$). Show that C_φ is cyclic if and only if $\arg \varphi'(\alpha)$ is an irrational multiple of π.

    (b) Suppose φ is loxodromic (hence conjugate to a complex dilation $z \to \lambda z$ where $|\lambda| < 1$). Show that C_φ is cyclic.

12. We saw in the second part of the Linear Fractional Hypercyclicity Theorem that if $\varphi \in LFT(U)$ is a parabolic non-automorphism, then for each $f \in H^2$ the only possible limit points of the C_φ-orbit of f are constant functions. Must there always be such a limit point? In other words, can an orbit be *discrete*? (Consider for example, $f(z) = \log(1 - z)$.)

13. *Parabolic orbit-convergence.* Show that the estimates involved in the proof of part (b) of the Linear Fractional Hypercyclicity Theorem imply that if φ is a parabolic non-automorphism, then $\sum(1 - |\varphi_n(p)|) = \infty$ for each $p \in U$. Show that, by contrast, $\sum(1 - |\varphi_n(p)|) < \infty$ for each $p \in U$ whenever φ is a parabolic *automorphism*.

14. *Chaotic linear operators.* For a continuous self-mapping T of a metric space X, consider the following definitions.

   - T has *sensitive dependence on initial conditions* if there exists $\delta > 0$ such that if $x \in X$ and $\varepsilon > 0$, then there exists a point Y in the ε-ball about x such that $\text{dist}(T^n x, T^n y) > \delta$ for some n.
   - A point $x \in X$ is called a *periodic point* of T if $T^n x = x$ for some positive integer n.
   - T is called *chaotic* if it is topologically transitive, has sensitive dependence on initial conditions, and has a dense set of periodic points.

Show that:

   (a) Every hypercyclic operator on a Banach space has sensitive dependence on initial conditions.
   (b) λB is chaotic on H^2 for every scalar λ of modulus > 1.
   (c) The backward shift B is *not* chaotic on the Bergman space A^2 (it is hypercyclic by Exercise 2 above).

7.5 Notes

Hypercyclicity. The Hypercyclicity Criterion of §7.1 was first proved by Carol Kitai in her Toronto dissertation [Kit '82]. This work, which contains many interesting results about hypercyclic operators, was never published, and the Criterion was rediscovered a few years later by Gethner and Shapiro [GrS '87], who used it to unify the results of Birkhoff and MacLane (Exercise 3 above), and to prove hypercyclicity for the backward shift on Bergman-type spaces (cf. Exercise 2).

Godefroy and Shapiro recently filled in everything between the theorems of Birkhoff and MacLane by proving: *If T is a continuous linear operator* *on $H(\mathbf{C})$ that commutes with differentiation and is not a scalar multiple of the identity operator, then T is hypercyclic on $H(\mathbf{C})$* [GyS '91].

Birkhoff's Theorem was recently generalized in another direction by Chan and Shapiro [ChS '91], who showed that translation operators are hypercyclic on Hilbert spaces of very slowly growing entire functions. By the Comparison Principle of §7.1, this provides for Birkhoff's original theorem hypercyclic vectors of very slow growth.

Rolewicz's Theorem appeared in [Rlz '69], and extensions to operators that commute with various generalizations of the backward shift are given in [GyS '91].

The definition of *chaotic* given in Exercise 14 above is due to Devaney, and occurs on page 50 of his book [Dvy '89].

Invariant subspaces and subsets. As mentioned at the beginning of this chapter, it is still not known if there is an operator on separable, infinite dimensional Hilbert space with no (closed, non-trivial) invariant subspace. Such operators do exist in the Banach space setting; the first example was constructed by Enflo [Enf '87], and later Read, after a couple of papers simplifying and extending Enflo's result, produced an operator on a separable Banach space for which every non-zero vector is hypercyclic [Rd '88]. Such an operator has no non-trivial closed invariant *subset*.

Cyclic and hypercyclic composition operators. The results of §7.2 and §7.3 are due to Bourdon and Shapiro [BoS '90, BoS '92], as are those of Exercise 11 of this chapter. Cyclicity for linear fractional composition operators on H^2 seems first to have been investigated by N. Zorboska [Zrb '87], who proved that the hyperbolic automorphisms induce cyclic composition operators.

In §7.1, for the proof that automorphisms with no interior fixed point induce hypercyclic composition operators, we needed to show that the polynomials that vanish at a fixed point outside the unit disc form a dense subset of H^2. This can also be regarded as a special case of Beurling's Theorem (see [Drn '70, §7.3]).

Hypercyclic perturbations of the identity. Exercise 4c above states that no perturbation of the identity by finite rank operators can be hypercyclic. This suggests that the same might be true for perturbations of the identity by *compact* operators, but Herrero and Wang [HeW '90] recently showed this is *false*. Subsequently Chan and Shapiro showed that the translation operators $f(z) \mapsto f(z + a)$, where $a \neq 0$, are hypercyclic perturbations of the identity by compacts when they act on small enough Hilbert spaces of entire functions. In fact the perturbing operator can be made "arbitrarily compact" by taking the space sufficiently small [ChS '91].

Universality. One can generalize the concept of hypercyclicity by using sequences of operators more general than the positive powers of a fixed operator. The resulting concept, often called "universality," has been studied for a long time in approximation theory, Fourier series, and complex analysis (see, for example, [Mcz '35] and [Mnf '47]). Further background and references can be found in [Ulv '72] and for a sampling of more recent work, along with further references, see [G-E '87], [Hzg '88], and [Luh '86].

Hypercyclic vectors. We observed early on that for a hypercyclic operator, "almost every" vector is hypercyclic. However the Hypercyclicity Criterion does not provide specific examples of hypercyclic vectors, and it appears that even in the simplest situations a concrete characterization of such vectors will never be within reach. On the other hand, Voronin [Vnn '75] has shown that the *Riemann zeta-function* has a certain hypercyclic-like property. Bourdon and Shapiro (unpublished) have shown that any hypercyclic vector for λB acting on H^2 ($|\lambda| > 1$) has hypercyclic outer factor. Bourdon recently showed that: *Every hypercyclic operator on Hilbert space has a (non-closed) dense invariant subspace that consists, except for the zero vector, entirely of hypercyclic vectors* [Bdn '92], thus completing earlier results that appeared in [Bmy '86], [GyS '91], and [Pvn '92].

8
Cyclicity and Models

The work of Chapter 6 on eigenvalues and Schröder's equation led us to the idea of representing univalent self-maps of U by simpler ones that act on more complicated domains, where the subtleties of the original map appear coded into the geometry of the new domain. In this chapter we extend our study of cyclicity by exploiting this idea to develop a method for transferring the linear fractional results of the last chapter to more general situations. In Chapter 9 we will accomplish something similar for the compactness problem.

8.1   Transference from Models

In this section we study univalent self-maps φ of the unit disc that are represented as

$$\varphi = \sigma^{-1} \circ \psi \circ \sigma \tag{1}$$

where $\psi \in LFT(U)$ has the same attractive fixed point as φ (the Denjoy-Wolff point if φ has no interior fixed point), and σ is a univalent map that takes U onto a simply connected domain $G = G(\varphi)$. When this happens, we say that σ *intertwines* φ and ψ, and call the pair (ψ, G) (or equivalently (ψ, σ)) a *linear fractional model* for φ. If, in addition, G is a *Jordan domain* (the region interior to a Jordan curve), we say (ψ, G) is a *Jordan model*. Our immediate goal is to show that in this situation the cyclic properties of C_ψ get "transferred" to C_φ. We will show that this method of transference applies to a large class of holomorphic self-maps φ of the unit disc.

For Jordan models, the Carathéodory Extension Theorem ([Rdn '87], §14.16–20) asserts that the "intertwining map" σ extends to a homeomorphism of the closed unit disc onto the closure of G, and equation (1) then yields an extension of φ to a homeomorphism of $\overline{U}$ onto the closure of $\varphi(U)$, so $\varphi(U)$ is also a Jordan domain. The commutative diagram below (Figure 8.1) illustrates the situation. To get a feeling for what is going to happen it helps to interpret this picture quite literally, taking for example:

- G to be a triangle in the unit disc with a vertex at the point $+1$,

- σ to be a univalent mapping of U onto G that fixes $+1$, and

- $\psi(z) = (1 + z)/2$.

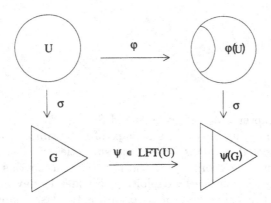

FIGURE 8.1. *A linear fractional model*

For lack of a better name, let us refer to this particular model as the *Standard Example*. Since $\psi(z) = (1 + z)/2$ is a hyperbolic self-map of U, the Linear Fractional Hypercyclicity Theorem of the last chapter asserts that it induces a hypercyclic composition operator on H^2. The next result—the key to all that follows—shows that this hypercyclicity is inherited by C_φ.

The Transference Theorem. *Suppose (ψ, G) is a Jordan model for φ, with $G \subset U$, and C_ψ hypercyclic (resp. cyclic) on H^2. Then C_φ is hypercyclic (resp. cyclic) on H^2.*

Proof. We prove only the hypercyclic version, the proof for the cyclic one being entirely similar. Suppose C_ψ is hypercyclic on H^2. Let $H^2(G)$ be the collection of functions $F \in H(G)$ for which $F \circ \sigma \in H^2$. For the remainder of the proof we keep track of the setting by writing $H^2(U)$ for H^2. We define the norm of a function F in $H^2(G)$ to be the $H^2(U)$ norm of $F \circ \sigma$. In other words we decree that the composition operator C_σ map $H^2(U)$ isometrically onto $H^2(G)$; in particular, $H^2(G)$ is a Hilbert space. By equation (1), C_σ establishes an isometric similarity between C_φ acting on $H^2(U)$ and C_ψ acting on $H^2(G)$. Thus to prove that C_φ is hypercyclic on $H^2(U)$, we need only show that the linear-fractional operator C_ψ is hypercyclic on $H^2(G)$.

By hypothesis C_ψ is hypercyclic on $H^2(U)$, so according to the Hypercyclic Comparison Principle of §7.1, we need only show that $H^2(G)$ has the following properties:

(a) $H^2(U) \subset H^2(G)$,

(b) $H^2(U)$ convergence $\Rightarrow H^2(G)$ convergence, and

(c) $H^2(U)$ is dense in $H^2(G)$.

For the first two properties, observe that because $G \subset U$, Littlewood's Theorem guarantees that C_σ is a bounded operator on $H^2(U)$. Suppose $f \in H^2(U)$. Then, denoting the norms in $H^2(G)$ and $H^2(U)$ respectively by $\| \ \|_G$ and $\| \ \|_U$, we have (upon identifying the function f with its restriction to G)

$$\|f\|_G \overset{\text{def}}{=} \|C_\sigma f\|_U \le \|C_\sigma\| \, \|f\|_U \ ,$$

which proves (a) and (b) at one stroke.

For property (c), we already know that the polynomials form a dense subset of $H^2(U)$, so by (b) above we will be finished if we can show that they are also dense in $H^2(G)$. A hopeful sign in this direction comes from Runge's Theorem [Rdn '87, Chapter 13], which shows that the polynomials are dense in $H(G)$, in the topology of uniform convergence on compact subsets of G. However the density of polynomials in $H^2(G)$ is a considerably stronger property, for whose proof we employ a deep result of Walsh about uniform holomorphic approximation. Let $A(G)$ denote the space of complex valued functions that are continuous on the closure of G and analytic on G. We equip $A(G)$ with the supremum norm. The result we need is:

Walsh's Theorem. *If G is a Jordan domain, then the polynomials are dense in $A(G)$.*

We will not prove Walsh's Theorem here; for a proof, see [Mkv '67, Theorem 3.9, page 98]. Nowadays it is most often regarded as a special case of:

Mergelyan's Theorem. *If K is a compact subset of the plane whose complement is connected, then every complex function that is continuous on K and analytic on its (topological) interior can be uniformly approximated on K by polynomials.*

For a proof of this result, see [Rdn '87, Chapter 20]. Returning to our argument, let $\mathcal{P}$ denote the collection of (holomorphic) polynomials. By Walsh's Theorem, $\mathcal{P}$ is dense in $A(G)$. Recall that, because G is a Jordan domain, σ extends to a homeomorphism of $\overline{U}$ onto $\overline{G}$ (Carathéodory's Theorem). Thus the operator C_σ also acts as an isometry of $A(G)$ onto $A(U)$, hence $C_\sigma(\mathcal{P})$ is dense in $A(U)$, which is in turn dense in $H^2(U)$ (because, for example, $A(U)$ contains the polynomials). Since the norm of $A(U)$ is larger than that of $H^2(U)$ (§1.2) we see that $C_\sigma(\mathcal{P})$ is dense in $H^2(U)$. Upon applying the inverse of C_σ, which is an isometry of $H^2(U)$ onto $H^2(G)$, we see that $\mathcal{P}$ is dense in $H^2(G)$. This completes the proof of hypercyclicity. $\square$

Remark. The Jordan property of G enters in this argument only to show that the polynomials are dense in $H^2(G)$. Thus the conclusion of the Transference Theorem holds as well for models (ψ, G), where G has this more general "H^2 polynomial approximation property." The subject of classifying the simply connected plane domains that have "mean approximation" properties like this is a fascinating one that has drawn considerable attention over the years. For more on this, see the *Notes* at the end of this chapter.

8.2   From Maps to Models

The partnership between the Riemann Mapping Theorem and the Transference Theorem makes it possible to construct cyclic and hypercyclic operators simply by drawing pictures of appropriate plane domains G. But suppose we *start* with a univalent self-map of U. How do we know if it comes from an "appropriate" picture? Does it come from any picture at all? Encouragingly, the answer to this last question is "yes."

The Linear Fractional Model Theorem. *Every univalent self-map of U has a linear fractional model.*

The proof of this beautiful result spans nine decades, and several authors (see *Notes* at the end of this chapter). It begins with Königs's 1884 theorem on the holomorphic solutions of Schröder's equation, that we discussed in Chapter 6. In our present terminology Königs's Theorem asserts:

> If φ is univalent and fixes the origin, then the pair $(\lambda z, \sigma)$ is a linear fractional model for φ, where $\lambda = \varphi'(0)$ and σ is the Königs function for φ.

If, on the other hand, φ fixes no point of U, then the Denjoy-Wolff Theorem (Chapter 5) asserts that there is an "attractive boundary fixed point" ω with $0 < \varphi'(\omega) \le 1$, where $\varphi'(\omega)$ denotes the angular derivative at ω.

For the case $\varphi'(\omega) = \lambda < 1$, Valiron showed that there is a univalent map σ of U such that the pair $(\lambda z + 1 - \lambda, \sigma)$ is a linear fractional model for φ [Vln '31]. In the course of proving the main result of the next section (a general hypercyclicity theorem), we will prove a version of Valiron's Theorem, where the hypotheses are strengthened, and correspondingly more information is obtained about the map σ.

However, we will prove neither Valiron's original result, nor the rest of the Linear Fractional Model Theorem. While the existence of a model (ψ, G) for a univalent self-map φ of U furnishes important moral support, it does not solve all our problems since, to satisfy the hypotheses of the Transference Theorem, the model domain G must also be a Jordan sub-domain of U.

This extra hypothesis is not to be taken lightly. For example, even if the original map φ takes U onto a Jordan domain, we have no assurance that this property is inherited by the corresponding "model domain" G. In fact, as shown by Exercise 2 below, it may well not be.

8.3   A General Hypercyclicity Theorem

In seeking to generalize the Linear Fractional Hypercyclicity Theorem of §7.2, keep in mind that any holomorphic self-map φ of U that hopes to induce a hypercyclic composition operator on H^2 must first satisfy some stringent necessary conditions. We have already seen in §7.2 that φ can have no fixed point in U. An entirely similar proof shows that φ must also be *univalent*, and with a bit more effort one can show that there has to be an additional "almost everywhere" univalence condition on the boundary function (see Exercise 8 below). For example, the conformal mapping of U onto the unit disc with the unit interval removed does not induce a hypercyclic operator.

We are going to work with a class of maps φ that satisfies these necessary conditions, along with a few additional regularity properties designed to make the analysis tractable. In the first place, we demand that the image of φ be a Jordan domain, so φ extends univalently to the *closed* unit disc. Next, we insure that the φ has no fixed point in U by forcing $\varphi(\overline{U})$ to contact ∂U at the point $+1$, and nowhere else. We further require that $\varphi(+1) = +1$, and that φ be differentiable at $+1$, with $\varphi'(1) \leq 1$. (Recall from Chapter 5 that if a holomorphic self-map of U has an angular derivative at a boundary fixed point, then this derivative must be positive.) The Grand Iteration Theorem of §5.1 now insures that $+1$ is the Denjoy-Wolff attracting point of φ, so in particular φ has no fixed point in U.

Finally we require, for technical reasons, that φ'' extend continuously to $U \cup \{1\}$, a condition which implies that φ' also extends continuously to $+1$. This extendability of the second derivative can be considerably weakened (see the *Notes* at the end of this chapter), but our methods will require something more than just extendability of the first derivative. Here is a summary of the hypotheses we will be placing on our maps:

(φ1) φ is continuous and univalent on $\overline{U}$.

(φ2) φ has $+1$ as a fixed point, with $\varphi'(+1) \leq 1$.

(φ3) The second derivative φ'' extends continuously to the point $+1$.

(φ4) $\varphi(\overline{U}) \subset U \cup \{+1\}$.

Having set the stage, we state the main result of this chapter.

Theorem. *Suppose φ satisfies hypotheses (φ 1) $-$ (φ 4) above.*

  (a) *If $\varphi'(1) < 1$, then C_φ is hypercyclic on H^2.*

  (b) *If $\varphi'(1) = 1$ and $\mathrm{Re}\,\varphi''(1) > 0$, then C_φ is not hypercyclic on H^2.*

To get some feeling for these hypotheses, recall from Chapter 0 that the hyperbolic members of $LFT(U)$ are those with derivative < 1 at the attracting fixed point ω, while the parabolic ones have derivative $= 1$ there. For parabolic maps with $\omega = +1$, the second derivative always has real part non-negative, and this real part is strictly positive if and only if the map is a non-automorphism. More generally, as will become evident in the proof of part (b):

> If φ is any holomorphic self-map of U that satisfies (φ2) and (φ3), and has $\varphi'(1) = 1$, then $\mathrm{Re}\,\varphi''(1) \geq 0$.

(In Exercise 4 below you will see that this is just another way of expressing the fact that the the boundary of $\varphi(U)$ has curvature ≥ 1 at $+1$.)

In particular, one expects that φ will have a model that is hyperbolic in part (a) of the result above, and parabolic non-automorphic in part (b). This is entirely correct, although only the proof of part (a) will require the model. Granting this, one can view the Theorem as asserting that whenever a map satisfies conditions (φ-1)–(φ-4), then its derivatives at the Denjoy-Wolff point determine the type of its model, and—at least in the hyperbolic and parabolic non-automorphism cases—the map has the same hypercyclic properties as the modelling map. It turns out, though we will not prove this, that under a little more differentiability at $+1$ the same is true in the parabolic automorphism case, and together these cases exhaust *all* possibilities (see *Notes* at the end of this section).

Proof of (a). Write $\lambda = \varphi'(1)$, and $\psi(z) = \lambda z + (1 - \lambda)$. We will show initially that φ has a Jordan model (ψ, G), where G lies in the half-plane $\{\mathrm{Re}\,z < 1\}$. Then a conformal transformation will take this model into one that satisfies all the hypotheses of the Transference Theorem.

We claim there is a univalent map σ that takes U onto a Jordan domain, with $\operatorname{Re}\sigma < 1$ and $\sigma \circ \varphi = \psi \circ \sigma$. For this we modify the method used to prove Königs's Theorem; to emphasize the similarity we use the map $z \mapsto 1 - z$ to transform the setting from U to the disc U_0 of radius one that is centered at $+1$. To keep the notation under control we retain the symbol φ to denote the corresponding resulting self-map of U_0. Thus $\varphi(0) = 0$, and both the first and second derivatives of φ extend continuously to the origin. These properties translate into the following "finite Taylor expansion" of φ at the origin:

$$\varphi(z) = \lambda z + z^2 B(z) \qquad (z \in \overline{U}_0) \tag{2}$$

where $0 < \lambda = \varphi'(0) < 1$, and the function B is bounded on the closure of U_0. The idea is to resurrect the proof of Königs's Theorem (§6.1), using an estimate derived from (2) in place of the Schwarz Lemma. We will obtain a solution σ of Schröder's equation $\sigma \circ \varphi = \lambda\sigma$ on U_0, where σ is now a Jordan map with positive real part.

We proceed, as in the original proof of Königs's Theorem, obtaining σ as a limit of normalized iterates $\sigma_n = \lambda^{-n}\varphi_n$. Note first that since φ has positive real part, so does each of its iterates, and therefore so does each map σ_n. We claim that the sequence $\{\sigma_n\}$ converges uniformly on $\overline{U}_0$.

For this, let $\beta = \max\{|B(z)| : z \in \overline{U}_0\}$, and let Δ be the intersection of $\overline{U}_0$ with the closed disc of radius $(1 - \lambda)/2\beta$ centered at the origin. Of course it may happen that $\beta \leq (1 - \lambda)/2$, in which case $\Delta = \overline{U}_0$. However if $\beta > (1 - \lambda)/2$, then the situation is the one shown in Figure 8.2 below.

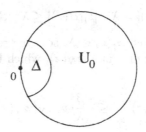

FIGURE 8.2. *The closed neighborhood* Δ *(for β large).*

In any case, by (2) above we have

$$|\varphi(z)| \leq \left(\frac{1+\lambda}{2}\right)|z| \tag{3}$$

for each $z \in \Delta$. Since $(1 + \lambda)/2 < 1$, this last inequality insures that $\varphi(\Delta) \subset \Delta$, so inequality (3) can be iterated for each $z \in \Delta$ to yield

$$|\varphi_n(z)| \leq \left(\frac{1+\lambda}{2}\right)|\varphi_{n-1}(z)|$$

$$\leq \left(\frac{1+\lambda}{2}\right)^2 |\varphi_{n-2}(z)|$$

$$\vdots$$

$$\leq \left(\frac{1+\lambda}{2}\right)^n |z|.$$

By our definition of Δ, this last estimate shows that

$$|\varphi_n(z)| \leq \frac{1-\lambda}{2\beta} \left(\frac{1+\lambda}{2}\right)^n \qquad (z \in \Delta) \qquad (4)$$

for each non-negative integer n. Note that the origin is now playing the role of Denjoy-Wolff point for φ, so $\varphi_n \to 0$ uniformly on compact subsets of U_0. We are assuming that φ takes $\overline{U}_0$ into $U_0 \cup \{0\}$, so the closure of $\varphi(\overline{U}_0)\backslash\Delta$ in U_0 is compact. Thus $\varphi_n \to 0$ *uniformly on $\overline{U}_0$*. In particular, there is a positive integer N such that

$$\varphi_n(\overline{U}_0) \subset \Delta \qquad (n \geq N).$$

We now proceed as in the original proof of Königs's Theorem, setting

$$F(z) = \frac{\varphi(z)}{\lambda z} \qquad (z \in \overline{U}_0),$$

and noting that for each $z \in \overline{U}_0$ the expansion (2) implies

$$|1 - F(z)| = \lambda^{-1}|z||B(z)| \leq \lambda^{-1}\beta|z|. \qquad (5)$$

Now fix $z \in \overline{U}_0$. If $j \geq N$ then $\varphi_j(z) \in \Delta$, so using respectively (5) and (4) above (with $j - N$ in place of n and $\varphi_N(z)$, which belongs to Δ, in place of z), we obtain

$$|1 - F(\varphi_j(z))| \leq \frac{\beta}{\lambda}|\varphi_{j-N}(\varphi_N(z))|$$

$$\leq \frac{1-\lambda}{2\lambda}\left(\frac{1+\lambda}{2}\right)^{j-N}$$

for each $z \in \overline{U}_0$. Since N is independent of the point $z \in \overline{U}_0$, this last inequality shows that each term of the infinite series $\sum |1 - F(\varphi_j(z))|$ is dominated by the corresponding term of a convergent geometric series. Thus $\sum |1 - F(\varphi_j(z))|$ converges uniformly on $\overline{U}_0$, by the Weierstrass M-test. Since this convergence is passed on to infinite product

$$\prod_{j=0}^{\infty} F(\varphi_j(z)) = z^{-1} \lim_{n\to\infty} \sigma_n(z),$$

the sequence $\{\sigma_n\}$ therefore converges uniformly on $\overline{U}_0$ to a function σ that fixes the origin, is continuous on $\overline{U}_0$, is holomorphic and univalent on U_0, and obeys Schröder's equation on $\overline{U}_0$. Furthermore, σ has positive real part on U_0 since it is non-constant there and, as we noted above, each σ_n has positive real part.

Thus, in order to show that $\sigma(\overline{U}_0)$ is a Jordan domain, it only remains to check that σ is univalent on ∂U. But this follows from Schröder's equation and the univalence of φ on $\overline{U}_0$. The argument is this: If $\sigma(z_1) = \sigma(z_2)$ for a pair of points $z_1, z_2 \in \partial U_0$, then upon multiplying by λ and using Schröder's equation (which we have just proved holds on the *closed* disc) we see that $\sigma(\varphi(z_1)) = \sigma(\varphi(z_2))$. If neither z_1 nor z_2 is zero, then both φ-images belong to U_0, on which we know σ is univalent. Thus $\varphi(z_1) = \varphi(z_2)$, so $z_1 = z_2$ since φ is assumed to be univalent on $\overline{U}_0$.

Suppose on the other hand that one of the original points, say z_1 is zero. Then Schröder's equation and the fact that $\sigma(0) = 0$ yield $\sigma(\varphi(z_2)) = 0$. But if $z_2 \neq 0$, then $\varphi(z_2) \in U_0$, contradicting the fact that $\operatorname{Re}\sigma > 0$ on U_0. Thus $z_2 = 0$, so σ is univalent on $\overline{U}_0$.

In summary, we have shown that there is a continuous, univalent map σ defined on $\overline{U}_0$ that has non-negative real part, is holomorphic on U_0, and satisfies Schröder's equation $\sigma \circ \varphi = \lambda\sigma$ on $\overline{U}_0$.

Upon transferring this result back to the unit disc by means of the map $z \mapsto 1 - z$ our accomplishment looks like this:

> If φ obeys the hypotheses of the Theorem, then it has a Jordan model (ψ, σ), where $\psi(z) = \lambda z + (1 - \lambda)$, and σ maps U into the half-plane $\{\operatorname{Re} z < 1\}$.

The only problem remaining is that G need not lie in U, but this is easily remedied. Let τ be a linear fractional transformation that takes the half-plane $\{\operatorname{Re} z < 1\}$ onto the unit disc, and fixes $+1$. Necessarily $\tau \in LFT(U)$. Define:

- $\tilde{\psi} = \tau \circ \psi \circ \tau^{-1}$, another member of $LFT(U)$ with attractive fixed point at $+1$,

- $\tilde{G} = \tau(G)$, a Jordan sub-domain of U, and

- $\tilde{\sigma} = \tau \circ \sigma$, a univalent mapping of U onto $\tilde{G}$.

Then $\tilde{\sigma} \circ \varphi = \tilde{\psi} \circ \tilde{\sigma}$, so $(\tilde{\psi}, \tilde{G})$ is a linear fractional model for φ. Since ψ is hyperbolic, so is $\tilde{\psi}$. The Linear Fractional Hypercyclicity Theorem of the last chapter now insures that the operator $C_{\tilde{\psi}}$ is hypercyclic on $H^2(U)$, hence by the Transference Theorem, the same is true of C_φ. □

We remark that the C^2 differentiability hypothesis on φ at the Denjoy-Wolff point can be weakened a bit; see Exercise 5 below.

Proof of (b). For convenience let $a = \varphi''(1)$; our assumption here is that $\operatorname{Re} a > 0$. The differentiability hypothesis on φ yields the following two-term Taylor expansion with remainder, centered at $+1$:

$$1 - \varphi(z) = (1 - z) - \left(\frac{a}{2} + \gamma(z)\right)(1 - z)^2 \qquad (z \in \overline{U}), \qquad (6)$$

where $\gamma(z) \to 0$ as $z \to 1$. Since the point $+1$ has been set up to be the Denjoy-Wolff attractor for φ, we know that $\lim_n \varphi_n(z) \to 1$ for every $z \in U$. The rest of the argument involves a detailed analysis of just how the orbits $\{\varphi_n(z)\}$ converge to $+1$. We break it into two parts.

PART I: *Non-tangential convergence of orbits.* We claim that for each $z \in U$ the orbit $\{\varphi_n(z)\}$ converges to $+1$ *non-tangentially*, in other words, it stays entirely within one of the lens-shaped regions L_α ($0 < \alpha < 1$) introduced in §2.3 (see Figure 2.1), where the angular opening $\alpha\pi/2$ depends, of course, on the initial point z.

The corresponding result for maps having angular derivative < 1 at the the Denjoy-Wolff point was treated in §5.2, and the proof here is quite similar. Once again, the situation is best understood by transferring the problem to the right half-plane Π, using the map $w = (1 + z)/(1 - z)$, which takes $+1$ to ∞ and the lens L_α to an angular sector $S_{\alpha\pi/2}$, where we recall the notation

$$S_\theta = \{w \in \Pi : |\arg w| < \theta\}.$$

We saw in Chapter 4 that this change of variable transforms the original map φ into a holomorphic self-map Φ of Π, with the relationship between the two maps expressed by the equations

$$1 - z = \frac{2}{w + 1} \quad \text{and} \quad 1 - \varphi(z) = \frac{2}{\Phi(w) + 1}.$$

When substituted into the Taylor expansion (6) these equations yield, after some routine calculation, the following expansion for Φ with "center at ∞":

$$\Phi(w) = w + a + \Gamma(w) \qquad (w \in \overline{\Pi}), \qquad (7)$$

where $\Gamma(w) \to 0$ as $w \to \infty$. Thus Φ is "essentially" translation by the vector $a = \varphi''(1)$, which we emphasize once again has strictly positive real part.

At this point we take time out to make good on an earlier promise. Note that the expansion (7) requires only that the original map φ obey the standing hypotheses $(\varphi 2)$ and $(\varphi 3)$ of this section, with $\varphi'(1) = 1$. Now the fact that Φ is required to take the right half-plane into itself dictates, as the reader can easily show, that $\operatorname{Re} a \geq 0$. Thus: *If φ obeys hypotheses $(\varphi 2)$ and $(\varphi 3)$, and $\varphi'(1) = 1$, then $\operatorname{Re} \varphi''(1) \geq 0$.* (Exercise 4 below provides another approach to this inequality.)

Returning to the proof of non-tangential convergence, since $\varphi_n(z) \to 1$ for each $z \in U$, we know that $\Phi_n(w) \to \infty$ for each $w \in \Pi$. We have to prove that for each such w there exists $0 < \beta < \pi/2$ such that $\{\Phi_n(w)\}_0^\infty \subset S_\beta$.

Fix $w \in \Pi$, and write $w_n = \Phi_n(w)$ for each positive integer n. Since $w_n \to \infty$ we know that for some positive integer N (which depends on w),

$$n > N \quad \Rightarrow \quad |\Gamma(w_n)| < \frac{1}{2}\operatorname{Re} a.$$

Now fix an integer $n > N$. By the last inequality and (7) we have

$$|w_{n+1} - w_n| < \frac{3}{2}\operatorname{Re} a \quad \text{and} \quad \operatorname{Re}(w_{n+1} - w_n) > \frac{1}{2}\operatorname{Re} a,$$

from which follows

$$\frac{|\operatorname{Im}(w_{n+1} - w_n)|}{\operatorname{Re}(w_{n+1} - w_n)} < 3\operatorname{Re} a.$$

Choose $0 < \theta < \pi/2$ so that $\tan\theta = 3\operatorname{Re} a$. The last inequality shows that

$$n > N \quad \Rightarrow \quad w_{n+1} - w_n \in S_\theta,$$

whereupon it follows upon summation (exactly as in the argument at the end of §5.2) that $w_n \in w_N + S_\theta$ for each $n > N$. Now choose $\theta < \beta < \pi/2$ so that $w_N + S_\theta \subset S_\beta$; then $w_n \in S_\beta$ for each $n > N$. This proves the desired non-tangential convergence. (In Exercise 7 below you will find that the orbit-convergence is *tangential* if $\operatorname{Re}\varphi''(1)$ is pure imaginary.)

PART II: *Estimates of functions on orbits.* This part of the proof is similar to the work done in the last chapter in proving that parabolic non-automorphisms induce non-hypercyclic operators. From the Taylor representation (6) of φ we know that

$$\lim_{z \to 1} \frac{1 - \varphi(z)}{1 - z} = 1 \quad \text{and} \quad \lim_{z \to 1} \frac{\varphi(z) - z}{(z - 1)^2} = \frac{a}{2}. \tag{8}$$

Now fix $z \in U$, and for notational convenience set $z_n = \varphi_n(z)$. Since $z_n \to 1$ non-tangentially as $n \to \infty$ we have for each n,

$$|1 - z_n| \leq \text{const.} \, (1 - |z_n|) \tag{9}$$

where here, and throughout the rest of the argument, the constant is independent of n. Upon substituting z_n for z in the second part of (8) and using (9) we obtain

$$|z_{n+1} - z_n| \leq \text{const.} \, (1 - |z_n|)^2. \tag{10}$$

Similarly, the same substitution in the first part of (8) provides, in conjunction with (9)

$$1 - |z_{n+1}| \geq \text{const.} \, (1 - |z_n|) \tag{11}$$

(an inequality which also follows from the Julia-Carathéodory Theorem).

Now fix $f \in H^2$, and use estimates (9)–(11), together with the "functional-difference" estimate (6) of §7.2 to obtain

$$|f(z_{n+1}) - f(z_n)| \leq \text{const.} \frac{|z_{n+1} - z_n|}{(\min\{1 - |z_n|, 1 - |z_{n+1}|\})^{3/2}}$$

$$\leq \text{const.} \frac{(1 - |z_n|)^2}{(1 - |z_n|)^{3/2}}$$

$$= \text{const.} (1 - |z_n|)^{1/2}$$

$$\to 0 \quad (n \to \infty).$$

To summarize: if $f \in H^2$ then

$$\lim_{n \to \infty} [f(\varphi_{n+1}(z)) - f(\varphi_n(z))] = 0 \tag{12}$$

for each $z \in U$. Now fix $f \in H^2$, to be regarded as a possible candidate for a hypercyclic vector. We claim that f is not elected. To this end, suppose $g \in H^2$ is a limit point of the C_φ-orbit of f, i.e., an H^2 limit point of some subsequence of $\{f \circ \varphi_n\}$. As in the last chapter, because H^2 convergence implies pointwise convergence, we have for some subsequence $n_k \nearrow \infty$,

$$g(z) = \lim_{k \to \infty} f(\varphi_{n_k}(z))$$

$$= \lim_{k \to \infty} f(\varphi_{n_k+1}(z)) \quad [\text{by (12)}]$$

$$= g(\varphi(z))$$

for each $z \in U$.

Thus any limit point of the C_φ-orbit of f is a fixed point of C_φ. Since C_φ is not the identity operator, not every point of H^2 is fixed (example: $g(z) \equiv z$), so f is not a hypercyclic vector. Since f was an arbitrary member of H^2, the operator C_φ is not hypercyclic. $\square$

Remarks on part (b) C_φ orbits. As in the corresponding linear fractional result, the strict positivity of $\text{Re}\, a$ comes into the proof of part (b) only to obtain the non-tangential convergence of φ-orbits. A more careful analysis of the orbits of φ shows that $\sum(1 - |\varphi_n(z)|) = \infty$ for each $z \in U$, and this in turn implies that the only fixed points of C_φ on H^2 are the constant functions (see Exercise 6 below). Thus part (b) of the last Theorem actually has the same conclusion as the corresponding part of the Linear Fractional Hypercyclicity Theorem: *Only constant functions can adhere to C_φ-orbits.*

Jordan Models. It can be shown that under slightly better than C^3 differentiability at the Denjoy-Wolff point, the maps φ of part (b) have Jordan models (ψ, G), where ψ is a parabolic non-automorphism. However in order to use the Transference Theorem, it is also necessary to have $G \subset U$, and this cannot always be arranged! The condition that determines when G can be placed in U, preserving the character of the model, is a relationship between the second and *third* derivative of φ at the Denjoy-Wolff point that can be expressed as an inequality on the *Schwarzian derivative* of φ at that point. Nevertheless, even in the exceptional cases it can still be shown that all these maps induce *cyclic* composition operators, but this requires a more delicate proof that does not appeal to the Transference Theorem. The details can be found in [BoS '93]; here we are content to simply summarize the full result:

Theorem. *Suppose φ is a holomorphic self-map of U that satisfies hypotheses $(\varphi 1)$–$(\varphi 4)$, with the differentiability in $(\varphi 3)$ strengthened to C^4 (or just $C^{3+\varepsilon}$). Then:*

(a) *φ has a Jordan model (ψ, G).*

(b) *The first and second derivatives of φ at the Denjoy-Wolff point determine the type of ψ.*

(c) *The cyclic behavior of C_φ on H^2 is the same as that of C_ψ.*

To emphasize the connections between model and mapping, we amplify the content of this result in the table below, which should be regarded as an extension of the table in the last chapter that summarized the corresponding linear fractional results. In the first two rows of the present table, asterisks (*) mark results proved in this chapter; the rest can be found in [BoS '93].

8.4 Exercises

1. Use the Transference Theorem to derive the hyperbolic non-automorphism part of the Linear Fractional Hypercyclicity Theorem of Chapter 7 from the corresponding automorphic result of §7.1.

2. *A Jordan map with non-Jordan model.* Let G be the crescent-shaped region obtained by omitting from the unit disc the closure of the left half of a lens L_α for some fixed $0 < \alpha < 1$. Show that for $0 < p < 1$ the map $\psi(z) = (p + z)/(1 + pz)$, which fixes the points ± 1, maps G into itself, and that the pair (ψ, G) is a linear fractional model for a univalent self-map φ of U, where $\varphi(U)$ is a Jordan domain.

Derivatives of φ at D.W. point. ($\omega = 1$)	$0 < \varphi'(1) < 1$	$\varphi'(1) = 1$ $\operatorname{Re} \varphi''(1) = 0$	$\varphi'(1) = 1$ $\operatorname{Re} \varphi''(1) > 0$
Cyclicity of C_φ	Hypercyclic*	Hypercyclic	Cyclic Not hypercyclic*
L.F. Model for φ	Hyperbolic*	Parabolic automorphism	Parabolic non-automorphism
Right $\frac{1}{2}$-plane rep'n. of model	$\Psi(w) = \lambda w + b$ $\lambda = \varphi'(1)$	$\Psi(w) = w + b$ $b = \varphi''(1)$	

Cyclic behavior, $\varphi \in \mathbf{C}^{3+\varepsilon}$ at $+1$ [BoS '93].

Suggestion: Look at the corresponding problem in the right half-plane.

3. *Non-uniqueness of models.* Suppose G is the region between the unit circle and the circle of radius $1/2$ centered at the point $1/2$. Let ψ be a parabolic automorphism of U with fixed point at $+1$. Show that (ψ, G) is a linear fractional model for a *hyperbolic automorphism* of U. (Note that this gives another way in which Jordan maps φ can have non-Jordan models).

4. Suppose φ is a holomorphic self-map of U that has a C^2 extension to ∂U, with $\varphi(1) = \varphi'(1) = 1$. Show that the curvature of $\varphi(\partial U)$ at $+1$ is
$$\kappa(1) = 1 + \operatorname{Re} \varphi''(1)$$
(cf. Exercise 6 of Chapter 0), and therefore $\operatorname{Re} \varphi''(1) \geq 0$.

5. Show that the conclusion of part (a) of the Hypercyclicity Theorem of §8.3 remains true if the standing hypothesis (φ-3) is replaced by the weaker assumption that $\varphi \in \mathbf{C}^{1+\varepsilon}$ at $+1$, for some $\varepsilon > 0$. Equivalently, assume that when the problem is transformed to $\overline{U}_0$, the expansion (2) is replaced by
$$|\varphi(z) - \lambda z| \leq \text{const. } |z|^{1+\varepsilon} \qquad (z \in \overline{U}_0).$$

6. Referring again to the same hypercyclicity theorem, show that the sum $\sum (1 - |\varphi_n(z)|)$ converges for the maps φ of part (a), and diverges for those of part (b). Show that the divergence result insures that for the maps φ of part (b), only constant functions can be limit points of C_φ-orbits (cf. the Linear Fractional Hypercyclicity Theorem of §7.2).

7. Suppose φ satisfies hypotheses ($\varphi 2$) and ($\varphi 3$) of §8.3 with $\omega = 1$, and that $\varphi'(1) = 1$ and $\varphi''(1) \neq 0$. Show that if $\operatorname{Re}\varphi''(1) = 0$ then for every $z \in U$ the orbit $\{\varphi_n(z)\}$ converges to $+1$ *tangentially* (Recall that in the course of proving the hypercyclicity theorem of §8.3 we showed that if $\operatorname{Re}\varphi''(1) \neq 0$ then this convergence is non-tangential).

8. Suppose φ is a holomorphic self-map of U that induces a hypercyclic composition operator on H^2. Show that:

   (a) φ is univalent on U.

   (b) φ is "univalent almost everywhere" on ∂U. By this we mean that there is a subset E of ∂U of Lebesgue measure zero such that the radial limit function of φ exists at every point of $\partial U \backslash E$ and is one-to-one on that set.

8.5 Notes

As mentioned in the text, the results of this chapter are from [BoS '93].

We observed at the end of §1 that the conclusion of the Transference Theorem holds under the weaker hypothesis that the holomorphic polynomials be dense in $H^2(G)$. Caughran proved that every *Carathéodory domain* (roughly speaking, a domain that has no "inner boundary points") has this approximation property ([Cgn '71]). In 1934 Farrell and Markushevich proved that for such a domain G, the polynomials are dense in the Bergman space $A^2(G)$, the Hilbert space of analytic functions on G that are square-integrable with respect to Lebesgue measure on G (see [Mkv '67, Theorem 3.20, page 117], and for further references [Gai '85, Theorem 1, page 17]). Recently, Bourdon [Bdn '87] proved for bounded, simply connected domains G: If the polynomials are dense in $A^2(G)$, then they are also dense $H^2(G)$. As Bourdon notes, Caughran's result follows quickly from this and the Farrell-Markushevich Theorem.

Exercise 2, the example of a Jordan map with non-Jordan model, is due to Carl Cowen. J. Akeroyd has investigated the H^2 polynomial approximation problem problem for "crescents" [Akd '87, Akd '92] (none of which are Carathéodory domains). He shows, for example, that this property fails for the domain G of Problem 2 above, whereas it holds for the region between two tangent circles. It does not seem to be known, however, if the mapping φ of Problem 2 induces a hypercyclic (or even cyclic) composition operator.

Such polynomial approximation problems play an important role in the study of cyclic behavior for other operators, most notably the operator of "multiplication by z" on various Hilbert spaces of analytic functions (see for example [AKS '91] and [Bdn '87] for some recent results, and [Shd '74] for an illuminating expository survey).

The Linear Fractional Model Theorem was developed over a period of about 90 years by Königs, Valiron, Baker, Pommerenke, and Cowen. It breaks down into three main cases, the first two of which have been already been noted in the text.

(a) φ has an interior fixed point, which we may without loss of generality assume is the origin. In this case, there is a model $(\lambda z, \sigma)$, where $\lambda = \varphi'(0)$ (Königs's Theorem).

If φ has no fixed point in U, recall from Chapter 5 that the Denjoy-Wolff attracting point point ω lies on ∂U, and $0 < \lambda = \varphi'(\omega) \leq 1$. For this situation:

(b) If $\lambda < 1$ then there is a model of the form $(\lambda z + (1 - \lambda), \sigma)$ ([Vln '31]).

(c) If $\lambda = 1$ then there is a model (ψ, σ) where $\psi \in LFT(U)$ is parabolic ([Pmk '79] and [BkP '79]).

The last case breaks into two subcases; ψ may be an automorphism, or not, and the characterization of these subcases requires considerable care. In this chapter we indicated without proof that under suitable differentiability hypotheses at the Denjoy-Wolff point, the second derivative distinguishes these subcases. But the general situation requires a difficult pseudohyperbolic separation condition on the points of orbits; for details see the original sources [Pmk '79] and [BkP '79].

Independently of Baker and Pommerenke, Cowen obtained the full theorem by a Riemann Surface construction [Cwn '81], and in the same paper also treated the problem of uniqueness of models (cf. Exercise 3). For further applications of models to problems about composition operators, see [Cwn '83] where they are used to study spectra, [CwK '88], where they are applied to questions about subnormality, and [SSS '92] (part of which is discussed in the next chapter), where they figure in the study of compactness.

Local versions of the different cases of the Linear Fractional Model Theorem have long been important in complex dynamics. In this setting one considers functions f holomorphic in a neighborhood of the origin, with $f(0) = 0$ and $f'(0) \in \overline{U}$. The case $0 < |\lambda| < 1$ corresponds to parts (a) and (b) of the Linear Fractional Model Theorem; and there results the "Local Königs's Theorem" of Chapter 6, Exercise 3. The case $\lambda = 1$ corresponds to case (c) of the Model Theorem. Here the local result, called "The Petal Theorem," asserts the existence of a finite number of regions, emanating from the origin and shaped roughly like the petals of a flower, each of which is invariant under the action of f, and on each of which f is conjugate to a translation (see [Brd '91], Theorem 6.5.7, page 122).

In general, much more remains to be done in finding precisely when the cyclicity and hypercyclicity of a composition operator can be inferred from the corresponding properties of the model of its inducing map. Here we have concentrated on the boundary fixed point case, but some results are also known about cyclicity in the interior fixed point case, and these make further interesting connections with questions about polynomial approximation in $H^2(G)$ (see [BoS '93]).

9
Compactness from Models

Linear fractional models emerged in the last chapter as tools for determining the cyclic behavior of composition operators. In this one they figure in the study of *compactness*. Recall from §5.5 that a necessary (but by no means sufficient) condition for a composition operator to be compact is that its inducing map have a fixed point in U. Thus the model associated with such a map must arise from Königs's solution to Schröder's equation. We introduced this model in Chapter 6, and showed that compactness imposes severe restrictions on the geometry of the Königs domain of φ (the image of U under the Königs function σ). In particular:

> If φ is univalent and C_φ is compact on H^2, then the Königs domain contains no sector.

Our goal here is to turn this result around, and show that for a natural class of maps φ: "no sectors implies compactness."

9.1   Review of Königs's Model

We assume for the remainder of this chapter that φ is a univalent self-map of U that fixes an interior point, which without loss of generality we may take to be the origin. Following Königs, we showed in §6.1 that there is a univalent map σ of U, with $\sigma(0) = 0$, that obeys Schröder's equation on U:

$$\sigma \circ \varphi = \lambda \sigma$$

where $\lambda = \varphi'(0)$. Moreover σ is unique up to multiplication by a constant; when normalized so that $\sigma'(0) = 1$, we call σ the *Königs function* of φ, and refer to its image $G = \sigma(U)$ as the *Königs domain* of φ. Recall that Schröder's equation asserts that the mapping M_λ of multiplication by λ takes G into itself, and establishes the conjugacy $\varphi = \sigma^{-1} \circ M_\lambda \circ \sigma$ between the action of φ on U and that of M_λ on G. In the language of the last chapter, the pair (M_λ, G) is a linear fractional model, henceforth called the *Königs model* for φ.

We seek to be able to recognize when C_φ is compact by looking solely at the geometry of its Königs domain G.

9.2 Motivation

For the rest of this chapter, unless otherwise specified, (M_λ, G) is the Königs's model for a univalent map φ of U that fixes the origin. We are aiming for results that have the flavor:

> If G is small then C_φ is compact.

Here is a very simple version of this principle, where the smallness assumption is the strongest possible.

Proposition. *If G is bounded then there is a positive integer n such that $\|\varphi_n\|_\infty < 1$ (hence C_φ^n is compact).*

Proof. Since σ is univalent and $\sigma(0) = 0$, the continuity of σ^{-1} provides a number $\delta > 0$ such that $|z| < 1/2$ whenever $|\sigma(z)| < \delta$. Since σ is assumed bounded on U, and $0 < |\lambda| < 1$, we have $\lambda^n \sigma(U) \subset \delta U$ for some positive integer n. Upon replacing φ by φ_n in Schröder's equation we obtain

$$\sigma(\varphi_n(U)) = \lambda^n \sigma(U) \subset \delta U,$$

hence our choice of δ implies that $\varphi_n(U) \subset U/2$, that is, $\|\varphi_n\|_\infty < 1$. Thus by the First Compactness Theorem (§2.2) we see that $C_\varphi^n = C_{\varphi_n}$ is compact.   □

For maps φ with *unbounded* Königs domain the situation is more complicated. For example, recall that the lens maps introduced in §2.3 induce compact composition operators. For each of these maps the Königs domain is a horizontal strip (see §6.2), which although unbounded, is in some sense "small." On the other hand, the example below shows that it is possible for φ to have a similarly shaped Königs domain, yet have *no* power of its induced composition operator compact.

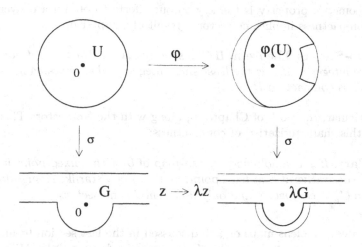

FIGURE 9.1. *G is "small," but no power of C_φ is compact.*

Example. Let G be the union of the strip $\{0 < \operatorname{Im} z < 1\}$ and the unit disc U, and fix any $0 < \lambda < 1$. Let φ be the map whose Königs model is (M_λ, G), as shown in Figure 9.1.
We claim that:

   No power of C_φ is compact on H^2.

Proof. Since the model for φ_n is just the model for φ with λ replaced by λ^n, it is enough to prove that C_φ itself is not compact. For this we need only observe that φ takes the arc γ of the unit circle that corresponds to the straight line $\{x > \lambda^{-1}\}$ in the lower boundary of G onto the arc that corresponds to the line $\{x > 1\}$. Thus $|\varphi(\zeta)| = 1$ for each $\zeta \in \gamma$, so it follows from the Proposition near the end of §2.5 that C_φ is not compact.   □

9.3   Main Result

In the last example the domain G is starlike with respect to the origin. G can, in fact, be made significantly smaller, still preserving its starlikeness and the non-compactness of C_φ^n, simply by replacing the upper boundary with a suitably shaped curve asymptotic to the lower boundary. The point is that the position of this lower boundary defeats any possibility of compactness for the powers of C_φ. To exclude this phenomenon, we adopt a stronger form of starlikeness.

Definition. G is *strictly starlike* if it contains the origin and $rw \in G$ whenever $0 \le r < 1$ and $w \in \overline{G}$ (here $\overline{G}$ denotes the closure of G in the complex plane).

This geometric property is the key to our efforts to obtain a converse to the "compactness implies no sectors" result of Chapter 6.

The No-Sectors Theorem. *If G is strictly starlike then $\lambda^n > 0$ for some positive integer n. If n is the least such integer, and G contains no sector, then C_φ^n is compact on H^2.*

In particular, the work of Chapter 6, along with the No-Sectors Theorem, yields this characterization of compactness:

Corollary. *If φ is a holomorphic self-map of U with a fixed point in U, a positive derivative at that fixed point, and a strictly starlike Königs domain G, then C_φ is compact if and only if G contains no sector.*

Examples. The lens maps of §2.5 discussed in the last section furnish examples of these results at work. More generally, for $\alpha > 0$ let G_α be the region described by the inequality $|y| < (|x| + 1)^\alpha$. Thus G_α is a strip if $\alpha = 0$, and for each $\alpha > 0$ it is a strictly starlike region that contains arbitrarily large discs. Let φ be the univalent self-map of U with Königs model (M_λ, G_α), where $0 < \lambda < 1$. Then the Corollary asserts that φ *is compact if and only if* $\alpha < 1$. In particular, compactness can arise from Königs domains that contain arbitrarily large discs (see Exercise 14 of §9.10 below for a more direct approach to the compactness of the resulting operators).

The proof of the No-Sectors Theorem requires estimates on the hyperbolic distance that will be developed in the next two sections. These estimates in turn provide preliminary "sector theorems" that are interesting in their own right. The final assault on the main theorem will require some results on radial limits (including the famous theorem of Lindelöf) which we develop in §9.8.

9.4   The Hyperbolic Distance on U

In Chapter 4 we made good use of the pseudo-hyperbolic distance $d(p, q)$ on the unit disc, a metric that induces the Euclidean topology on the disc, but sees conformal automorphisms as isometries. Our work on the No-Sectors Theorem will require something similar for the Königs domain G. An obvious strategy might be to use the map σ to transfer the pseudo-hyperbolic distance from U to $G = \sigma(U)$:

$$d_G(\sigma(p), \sigma(q)) \overset{\text{def}}{=} d(p, q) \qquad (p, q \in U).$$

The quantity so defined does indeed have properties analogous to those of the original pseudo-hyperbolic distance on U, but has the disadvantage of being difficult to relate to the Euclidean geometry of G. To get around this problem we are going to regroup and consider a different approach that produces a conformally invariant metric more amenable to Euclidean comparisons.

Definition. Suppose p and q are points of U and $\gamma : [a, b] \rightarrow U$ is a piecewise C^1 function with $\gamma(a) = p$ and $\gamma(b) = q$. We view γ as a curve in U with endpoints p and q, and define the *hyperbolic length* of γ to be:

$$\ell_U(\gamma) \overset{\text{def}}{=} \int_\gamma \frac{2|dz|}{1 - |z|^2} = 2 \int_a^b \frac{|\gamma'(t)|\, dt}{1 - |\gamma(t)|^2}.$$

The change-of-variable formula shows that this definition does not depend on the way the curve γ is parameterized. We define the *hyperbolic distance* from p to q to be the "length of the shortest curve from p to q," that is:

$$\rho_U(p, q) \overset{\text{def}}{=} \inf_\gamma \ell_U(\gamma),$$

where on the right, γ runs through all piecewise C^1 curves from p to q, and we use "infimum" instead of "minimum" because we don't yet know if there really *is* a shortest curve from p to q (we will see before long that a unique such curve does exist).

It is easy to check that the hyperbolic distance is a *metric* on U, and that it induces the usual Euclidean topology (see Exercise 3 of §9.10).

Example. For $0 < x < 1$,

$$\rho_U(0, x) = 2 \int_0^x \frac{dt}{1 - t^2} = \log \frac{1 + x}{1 - x}.$$

Proof. The integral, whose evaluation is elementary, is the hyperbolic length of the interval $[0, x]$, so the proof comes down to showing that this interval is hyperbolically the shortest path from 0 to x. To this end, suppose γ is any piecewise C^1 curve in U from 0 to x. Let u and v denote respectively the real and imaginary parts of the function γ. Then

$$\ell_U(\gamma) \quad = \quad 2 \int_a^b \frac{|\gamma'(t)|\, dt}{1 - |\gamma(t)|^2}$$

$$\geq \quad 2 \int_a^b \frac{|u'(t)|\, dt}{1 - |u(t)|^2}$$

$$\geq \quad 2 \int_0^x \frac{ds}{1 - s^2}$$

$$= \quad \ell_U[0, x],$$

where the second inequality reflects the fact that the function $u(t)$ may not be one-to-one. This shows that the segment from 0 to x is indeed the hyperbolically shortest path between the two points, and completes the proof.   □

Remarks. (a) In the above proof we referred to the interval $[0, x]$ as *the* hyperbolically shortest path from 0 to x, thus suggesting that it is the unique such path. This is indeed the case, as can be seen by examining the case of equalities in the two inequalities involved in the last estimate. There is equality in the first one if and only if $v \equiv 0$, and in the second one if and only if u is a one-to-one mapping of $[0, x]$ onto itself. Thus:

> The interval $[0, x]$ is the unique piecewise C^1 curve of shortest length between 0 and x.

(b) In the definition of hyperbolic length, the factor "2" in the numerator of the integrand is purely stylistic—any other positive constant will do as well, and many authors use "1" instead. The idea is that we are prescribing at each point $z \in U$ a "distortion function"

$$h_U(z) = \frac{2}{1 - |z|^2}$$

that alters the Euclidean distance in a way that will (hopefully) render the new distance conformally invariant. That it works is the subject of our next result. Conversely, Exercise 9 of §9.10 shows that h_U is, up to multiplication by a positive constant, the *only* distortion function that works.

We now establish the conformal invariance of hyperbolic distance.

Proposition. *Suppose* $\varphi \in \text{Aut}(U)$.

(a) *If* γ *is a piecewise* C^1 *curve in* U *then* $\ell_U(\varphi(\gamma)) = \ell_U(\gamma)$.

(b) *If* $p, q \in U$ *then* $\rho_U(\varphi(p), \varphi(q)) = \rho_U(p, q)$.

Proof. Clearly (b) follows from (a). The key to (a) is the fact that if $\alpha \in \text{Aut}(U)$ then for $z \in U$,

$$|\alpha'(z)| = \frac{1 - |\alpha(z)|^2}{1 - |z|^2}. \tag{1}$$

For $\alpha = \alpha_p$, the special automorphism that interchanges the origin with the point p (see §0.4), this is a straightforward computation (cf.§4.8 Exercise 5). The general case then follows from this one and the fact that each automorphism can be written as $\omega\alpha_p$, where $\omega \in \partial U$ and $p \in U$ (Exercise

2 of §0.5; see also [Rdn '87, §12.5]). With this formula in hand, part (a) follows upon changing variables in the integral for hyperbolic length:

$$\ell_U(\alpha(\gamma)) = 2 \int_{\alpha(\gamma)} \frac{|dw|}{1 - |w|^2}$$

$$= 2 \int_\gamma |\alpha'(z)| \frac{|dz|}{1 - |\alpha(z)|^2} \qquad [w = \alpha(z)]$$

$$= 2 \int_\gamma \frac{|dz|}{1 - |z|^2}$$

$$= \ell_U(\gamma)$$

where the third line follows from equation (1) above. □

A few paragraphs ago we computed the hyperbolic distance from the origin to a point x on the unit interval. The result can be rewritten as

$$\rho_U(0, x) = \log \frac{1 + d(0, x)}{1 - d(0, x)} \qquad (0 \le x < 1) \tag{2}$$

where d denotes the pseudo-hyperbolic distance on U. Since any pair of points in U can be moved to the pair $(0, x)$ by an automorphism, this formula and the conformal invariance of both pseudo-hyperbolic and hyperbolic distances yield:

Theorem. *For $p, q \in U$ we have*

$$\rho_U(p, q) = \ell_U(\gamma) = \log \frac{1 + d(p, q)}{1 - d(p, q)} \tag{3}$$

where γ is the unique arc that joins p and q, and lies on a circle perpendicular to the unit circle.

It is easy to see that the interval $[0, x]$ is the unique (hyperbolically) shortest path from 0 to x, so there is corresponding uniqueness for the arc γ of the Theorem. We call γ the *hyperbolic geodesic* joining p and q.

9.5 The Hyperbolic Distance on G

In this section G can be any simply connected plane domain not the whole plane, and σ is a univalent map taking U onto G. We use σ to transfer the notions of hyperbolic length and distance from U to G. More precisely, if Γ is a piecewise C^1 curve in G, then $\gamma = \sigma^{-1}(\Gamma)$ is another such curve in U, and we define the hyperbolic length of Γ by the equation $\ell_G(\Gamma) = \ell_U(\gamma)$. Similarly we define hyperbolic distance on G by:

$$\rho_G(\sigma(p), \sigma(q)) = \rho_U(p, q) \qquad (p, q \in U).$$

Thus, by decree, ℓ_G and ρ_G are invariant under the action of conformal automorphisms of G, and ρ_G is a metric on G that induces the usual Euclidean topology (see Exercise 3 of §9.10 below). Moreover

$$\rho_G(P, Q) = \inf_\Gamma \ell_G(\Gamma)$$

where Γ runs through all piecewise C^1 curves in G from P to Q, with the infimum attained uniquely on the σ-image of the hyperbolic geodesic in U that joins $\sigma^{-1}(P)$ to $\sigma^{-1}(Q)$. Finally, if Γ is a piecewise C^1 curve in G then the change of variable $w = \sigma(z)$ in the integral formula for the hyperbolic length of curves in U yields

$$\ell_G(\Gamma) = \int_\Gamma h_G(w) |dw| \tag{4}$$

where

$$h_G(\sigma(z)) = \frac{2}{(1 - |z|^2)|\sigma'(z)|} \qquad (z \in U). \tag{5}$$

The link between hyperbolic and Euclidean geometry on G is provided by these formulas, and a famous distortion theorem for univalent functions.

The Koebe One-Quarter Theorem. *Suppose f is any univalent function on U with $f(0) = 0$ and $|f'(0)| = 1$. Then*

$$f(U) \supset \frac{1}{4}U.$$

The constant $\frac{1}{4}$ is sharp, as can be seen by considering the image of the unit disc under the univalent mapping $f(z) = z/(1 - z)^2$. For our purposes $\frac{1}{4}$ can be replaced by any positive constant, so in the interest of economy we will employ the theorem as stated above, but will prove only the corresponding "$\frac{1}{24}$-version." For a proof of the full theorem, see [Rdn '87, Theorem 14.14, page 307]. The proof below comes from [Nvl '53, Ch. IV, §3.2, pp. 86–87], and in this reference you can also find a refinement that yields the sharp result (§3.4, pp. 88–90).

Proof. Write $G = f(U)$ and let δ denote the distance from the origin to ∂G. Thus $\delta > 0$ (since $f(0) = 0$), G contains the disc δU, and since this disc is the *largest* one centered at the origin and contained in G, some point of its boundary also lies on ∂G. Since we are assuming $|f'(0)| = 1$ (rather than the more traditional normalization $f'(0) = 1$), we may assume, upon suitably rotating the image domain if necessary, that this point is δ itself, i.e., that $\delta \in \partial G$.

Since the holomorphic function $w \mapsto \delta - w$ does not vanish anywhere on the simply connected domain G, it has a holomorphic square root h such that $h(0) = \sqrt{\delta}$ (positive square root). Thus h coincides on δU with the principal branch of the square root of $\delta - w$, and since the latter function has positive real part on δU, so does h. So $h(\delta U)$ lies in the right half-plane.

A map like h is used in the proof of the Riemann Mapping Theorem to reduce attention to the bounded case, and, just as in that proof, the key is that $h(G)$ is disjoint from $-h(G)$. (Otherwise there would be points $w, w' \in G$ such that $h(w) = -h(w')$, so upon squaring both sides, $w = w'$, whereupon $h(w) = 0$. This would force $w = \delta$, in contradiction to the assumption that $w \in G$.) In particular, $h(G)$ is disjoint from $-h(\delta U)$, a region which lies in the *left* half-plane, and contains the point $-\sqrt{\delta}$ (see Figure 9.2 below).

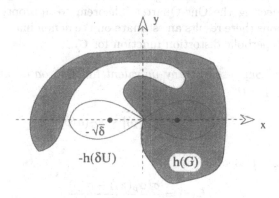

FIGURE 9.2. $h(G)$ disjoint from $-h(\delta U)$.

Thus, the mapping

$$\tau(w) = \frac{\sqrt{\delta} - w}{\sqrt{\delta} + w}$$

takes $h(G)$, and even the larger region $\widehat{\mathbf{C}} \backslash \{-h(\delta U)\}$, into a disc whose radius is, by the Maximum Principle,

$$\max_{w \notin -h(\delta U)} |\tau(w)| = \max_{\theta \in \mathbf{R}} \left| \frac{\sqrt{\delta} + h(\delta e^{i\theta})}{\sqrt{\delta} - h(\delta e^{i\theta})} \right|$$

$$= \max_{\theta} \frac{|\sqrt{\delta} + h(\delta e^{i\theta})|^2}{|\delta - (\delta - \delta e^{i\theta})|}$$

$$= (1 + \sqrt{2})^2 < 6.$$

The map $g = \frac{1}{6} \cdot \tau \circ h \circ f$ therefore takes the unit disc into itself, and since it also fixes the origin, the Schwarz Lemma demands that $|g'(0)| \leq 1$. On the other hand, a computation involving the chain rule shows that $|g'(0)| = (24\delta)^{-1}$. Thus $\delta \geq 1/24$, as desired.   □

The above work suggests that for each $P \in G$ we focus on the quantity

$$\delta_G(P) = \text{Euclidean distance from } P \text{ to } \partial G,$$

which is the Euclidean radius of the largest open disc in G with Euclidean center at P. In this language the One-Quarter Theorem asserts:

If σ is a "normalized" univalent mapping of U onto G then $\delta_G(0) \geq 1/4$.

(Note once again that the proof given here showed $\delta_G(0) \geq 1/24$).

Upon subjecting the One-Quarter Theorem to appropriate conformal transformations there results an estimate on the denominator in expression (5) for the hyperbolic distortion function for G.

Corollary 1. *Suppose σ is any univalent function on U, and $G = \sigma(U)$. Then for any $p \in U$*

$$\frac{1}{4}(1 - |p|^2)|\sigma'(p)| \leq \delta_G(\sigma(p)).$$

Proof. For $z \in U$ define

$$f_p(z) = \frac{\sigma(\alpha_p(z)) - \sigma(p)}{(1 - |p|^2)|\sigma'(p)|},$$

where α_p is the special automorphism that interchanges the point $p \in U$ with the origin. Thus f_p (often called the *Koebe transform* of f) is holomorphic and univalent on U, vanishes at the origin, and $|f'_p(0)| = 1$. By the One-Quarter Theorem, $f_p(U) \supset \frac{1}{4}U$, that is:

$$\sigma(U) \supset \sigma(p) + \left\{ \frac{1}{4}(1 - |p|^2)|\sigma'(p)| \right\} U,$$

and this is a restatement of the desired inequality.   □

Remark. This corollary is one half of a result commonly known as the Koebe Distortion Theorem. The other half is the upper estimate

$$\delta_G(\sigma(p)) \leq (1 - |p|^2)|\sigma'(p)| \qquad (\varphi \in U),$$

which follows quickly from the Schwarz Lemma (see Exercise 11 of §9.10 below).

In conjunction with equations (4) and (5) above, Corollary 1 immediately provides a crucial connection between the hyperbolic and Euclidean distances on G.

Corollary 2. *Suppose G is any simply connected domain, not the whole complex plane. Then for any piecewise C^1 curve Γ in G,*

$$\ell_G(\Gamma) \geq \frac{1}{2} \int_\Gamma \frac{|dw|}{\delta_G(w)}.$$

Remark. Taking into account the upper estimate on δ_G noted in the *Remark* above, we obtain the full story: for points $w \in G$,

$$\frac{1}{2\delta_G(w)} \leq h_G(w) \leq \frac{2}{\delta_G(w)},$$

hence for curves Γ in G,

$$\frac{1}{2} \int_\Gamma \frac{|dw|}{\delta_G(w)} \leq \ell_G(\Gamma) \leq 2 \int_\Gamma \frac{|dw|}{\delta_G(w)}.$$

Corollary 2 leads to an estimate of how the hyperbolic distance increases relative to the Euclidean as one moves toward the boundary.

Distance Lemma. *If G is a simply connected domain, not the whole complex plane, then for $P, Q \in G$,*

$$\rho_G(P, Q) \geq \frac{1}{2} \log \left(1 + \frac{|P - Q|}{\min\{\delta_G(P), \delta_G(Q)\}} \right).$$

Proof. We may suppose that P is no farther from ∂G than Q, i.c., $\delta_G(P) \leq \delta_G(Q)$. Let Γ be any curve in G from P to Q. By Corollary 2,

$$\ell_G(\Gamma) = \int_\Gamma h_G(w) \, |dw|$$

$$\geq \frac{1}{2} \int_\Gamma \frac{|dw|}{\delta_G(w)}.$$

For $w \in \Gamma$ let $s = s(w)$ denote the Euclidean arc length from P to w. Then by the (Euclidean) triangle inequality,

$$\delta_G(w) \leq \delta_G(P) + |P - w| \leq \delta_G(P) + s$$

which, along with the estimate of Corollary 2 yields

$$\ell_G(\Gamma) \;\geq\; \frac{1}{2}\int_{s=0}^{|\Gamma|} \frac{ds}{\delta_G(P)+s}$$

$$= \;\frac{1}{2}\log\left\{1 + \frac{|\Gamma|}{\delta_G(P)}\right\}$$

$$\geq \;\frac{1}{2}\log\left\{1 + \frac{|P-Q|}{\delta_G(P)}\right\}$$

where $|\Gamma|$ denotes the Euclidean length of Γ. The desired inequality now follows upon taking the infimum of the left hand side of the inequality above over all piecewise C^1 curves Γ in G from P to Q (or, equally well, by initially taking Γ to be the geodesic joining these two points). $\qquad\square$

To see an example of the Distance Estimate at work, consider the following consequence, which we will need in the next section.

Corollary. *Suppose $\{P_n\}$ is a sequence in G and $P_n \to P \in \partial G$. If $\{Q_n\}$ is another sequence in G with $\sup_n \rho_G(P_n, Q_n) < \infty$, then also $Q_n \to P$.*

Proof. Let $M = \sup_n \rho_G(P_n, Q_n)$. Then by the Distance Lemma and the fact that $P_n \to \partial G$,

$$|P_n - Q_n| \leq (e^{2M}-1)\delta_G(P_n) \to 0$$

hence $Q_n \to P$. $\qquad\square$

Note that for $G = U$ this result follows immediately from formula (3) that expresses the relationship between the pseudo-hyperbolic and hyperbolic distances.

9.6 Twisted Sectors

By now we have in hand enough hyperbolic technology to establish a major component of the proof of the No-Sectors Theorem. This result, which is of interest in its own right, gives a sufficient condition for a Königs domain to contain a "twisted version" of a sector.

Definition. Suppose Γ is an simple curve in the plane that joins the origin to ∞. For $\varepsilon > 0$, the *twisted sector about Γ of aperture ε* is the set

$$\mathcal{S}_\varepsilon(\Gamma) = \{z : |z - w| < \varepsilon|w|, w \in \Gamma\}.$$

Thus $\mathcal{S}_\varepsilon(\Gamma)$ is the set swept out by an open disc centered at w, of radius $\varepsilon|w|$ as its center w traverses the curve Γ. Any twisted sector about a ray (a line with one endpoint at the origin and the other at ∞) is, of course, an ordinary sector.

We now return to our standard setup, where φ is a univalent self-map of U that is not an automorphism and fixes the origin, σ is the Königs function of φ, and $G = \sigma(U)$ is the Königs domain. Recall that $w \in \partial U$ is called a *boundary fixed point* of φ if $\lim_{r \to 1-} \varphi(rw) = w$.

The Twisted Sector Theorem. *Suppose φ has an angular derivative at a boundary fixed point w. Then the curve*

$$\Gamma = \{\sigma(rw) : 0 \leq r < 1\}$$

is unbounded, and G contains a twisted sector about Γ.

Proof. We may suppose that $w = +1$, so $\varphi(r) \to 1$ as $r \to 1-$. We are assuming that φ has an angular derivative at $+1$, so,

$$\lim_{r \to 1-} \frac{1 - \varphi(r)}{1 - r} = \varphi'(1) \tag{6}$$

hence by the Julia-Carathéodory Theorem (§4.2)

$$\lim_{r \to 1-} \frac{1 - |\varphi(r)|}{1 - r} = \varphi'(1), \tag{7}$$

and $\varphi'(1) \in (1, \infty)$ (the fact that $\varphi'(1) \neq 1$ is not essential for this proof—it follows from the "Grand Iteration Theorem" of Chapter 5).

We claim that

$$\lim_{r \to 1-} \rho_U(r, \varphi(r)) = \log \varphi'(1) . \tag{8}$$

By equation (3) of §9.4, which relates the hyperbolic and pseudo-hyperbolic distances on U, the desired result (8) is equivalent to

$$\lim_{r \to 1-} d(r, \varphi(r)) = \frac{\varphi'(1) - 1}{\varphi'(1) + 1} . \tag{9}$$

Now recall formula (3) of §4.3 for the pseudo-hyperbolic metric:

$$1 - d(p, q)^2 = \frac{(1 - |p|^2)(1 - |q|^2)}{|1 - \bar{p}q|^2} .$$

Upon substituting r for p and $\varphi(r)$ for q in this formula, and doing a little algebra there results:

$$1 - d(r, \varphi(r))^2 = (1 + r)^2 \left| 1 + r\frac{1 - \varphi(r)}{1 - r} \right|^{-2} \frac{1 - |\varphi(r)|^2}{1 - r^2} .$$

Because of (6), (7), and the fact that $\lim_{r\to 1-} \varphi(r) = 1$, the expression on the right converges to $4\varphi'(1)(1 + \varphi'(1))^{-2}$ as $r \to 1-$. This yields (9), and therefore (8).

Since φ and σ obey Schröder's equation $\sigma \circ \varphi = \lambda\sigma$ (where $\lambda = \varphi'(0)$), we can transfer (8) to G as follows:

$$\rho_U(r, \varphi(r)) = \rho_G(\sigma(r), \sigma(\varphi(r))) = \rho_G(\sigma(r), \lambda\sigma(r)),$$

hence

$$\lim_{r\to 1-} \rho_G(\sigma(r), \lambda\sigma(r)) = \log\varphi'(1). \tag{10}$$

We claim that $\sigma(r) \to \infty$ as $r \to 1-$. Indeed, if this is not the case then there is a sequence $0 \le r_n \nearrow 1$ and a point $P \in \overline{G}$ such that $\sigma(r_n) \to P$. The univalence of σ insures that $P \notin G$, so we must have $P \in \partial G$. But (10) above shows that the quantity $\rho_G(\sigma(r_n), \lambda\sigma(r_n))$ remains bounded as $n \to \infty$, hence by the Distance Lemma of the last section,

$$0 = \lim_n |\lambda\sigma(r_n) - \sigma(r_n)| = |1 - \lambda||P|.$$

Since $P \neq 0$ (P is in the boundary of G, but $0 \in G$) we must have $\lambda = 1$, hence φ is the identity map on U, which we assumed at the outset is not the case. Thus $\sigma(r) \to \infty$.

To obtain the existence of a twisted sector in G about $\sigma([0,1))$ we use (10) above, along with the Distance Lemma to obtain

$$\frac{1}{2} \limsup_{r\to 1-} \log\left\{ \frac{|\sigma(r) - \lambda\sigma(r)|}{\delta_G(\sigma(r))} \right\} \le \limsup_{r\to 1-} \rho_G(\sigma(r), \lambda\sigma(r)) \le \log\varphi'(1),$$

whereupon

$$\limsup_{r\to 1-} \frac{|\sigma(r)|}{\delta_G(\sigma(r))} \le \frac{\varphi'(1)^2}{|1 - \lambda|}$$

which yields the desired conclusion.   □

9.7   Main Theorem: Down Payment

The results at hand lead quickly to a version of the No-Sectors Theorem where symmetry replaces strict starlikeness.

Theorem. *Suppose φ is a univalent self-map of U with $\varphi(0) = 0$ and $\varphi(U)$ symmetric about the real axis. Suppose further that the closure of $\varphi(U)$ intersects the unit circle uniquely at the point $+1$. Then the following statements are equivalent:*

(a) C_φ *is compact on H^2.*

(b) φ *does not have an angular derivative at $+1$.*

(c) $\sigma(U)$ *contains no sector.*

Proof. Recall that the existence of the angular derivative at $+1$ brings with it the requirement that $\varphi(r) \to 1$ as $r \to 1-$. Thus $+1$ is the *only* point of the boundary at which φ has a chance to have an angular derivative, so the equivalence of (a) and (b) above follows directly from the angular derivative criterion of Chapter 4. The fact that (a) (or (b)) implies (c) was done in Chapter 6, so it remains to prove that (c) implies (b).

Suppose (b) fails, i.e. that φ has an angular derivative at $+1$. Thus $\varphi(r)$ converges to some point of ∂U as $r \to 1-$. Since this point belongs to both the boundary of U and the closure of $\varphi(U)$, it must be $+1$. Thus $+1$ is a boundary fixed point of φ, and since φ has an angular derivative there, the Twisted Sector Theorem insures that $\sigma(r) \to \infty$ and the Königs domain G contains a twisted sector S about $\sigma([0,1))$.

By the reflection principle, the symmetry of $\varphi(U)$ forces φ to be real on the real axis, hence $\varphi([0,1)) = [0,1)$. In particular, $\lambda = \varphi'(0) > 0$. In Chapter 6 we showed that $\sigma = \lim \lambda^{-n}\varphi_n$, where φ_n is the n-th iterate of φ. Since each φ_n is positive on $[0,1)$, so is σ, hence $\sigma([0,1)) = [0,\infty)$. Thus, the twisted sector S is a *bona fide* sector in G, as desired.   □

9.8 Three Lemmas

Before beginning the final assault on the No-Sectors Theorem, we pause to collect three auxiliary results that are necessary for the argument.

Lemma 1 (Lines). *Suppose G is an unbounded strictly starlike domain, and $\{w_n\}$ is a sequence of points in G such that $|w_n| \to \infty$ and $\arg w_n \to \alpha$ as $n \to \infty$. Then the ray*

$$L_\alpha = \{te^{i\alpha} : t \geq 0\}$$

lies entirely in G.

Proof. Write $w_n = R_n e^{i\alpha_n}$ where $R_n > 0$ and $\alpha_n = \arg w_n \to \alpha$. Fix $0 < R < \infty$. Then for all sufficiently large n we have $R_n > R$, so by the starlikeness of G,

$$Re^{i\alpha_n} = \frac{R}{R_n}w_n \in G.$$

Thus $w = \lim_n Re^{i\alpha_n} \in \overline{G}$, so $L_\alpha \subset \overline{G}$. Therefore

$$L_\alpha = \frac{1}{2}L_\alpha \subset \frac{1}{2}\overline{G} \subset G$$

where the last inclusion follows from the *strict* starlikeness of G.   □

Corollary. *Every strictly starlike unbounded domain contains a ray.*

Proof. Since G is unbounded it contains a sequence $\{w_n\}$ with $|w_n| \to \infty$. By passing to a subsequence we may also assume that $\lim_n \arg w_n$ exists. The desired result now follows immediately from the Lemma. $\square$

Lemma 2 (Limits). *Suppose σ is a univalent function on U and $G = \sigma(U)$. Then:*

(a) *For almost every $\zeta \in \partial U$ the (possibly infinite) radial limit*

$$\sigma^*(\zeta) = \lim_{r \to 1-} \sigma(r\zeta)$$

exists.

(b) *σ^* is not a.e. constant on any subarc of ∂U.*

(c) *If $\Gamma : [0, \infty) \to G$ is any curve with*

$$\lim_{t \to \infty} \Gamma(t) = \infty,$$

then there exists $\omega \in \partial U$ such that

$$\lim_{t \to \infty} \sigma^{-1}(\Gamma(t)) = \omega.$$

Proof. (a) and (b): Just as in the proof of the One-Quarter Theorem, let Q be a point of the plane not in G, and let h be a holomorphic branch of the square root of $w \mapsto w - Q$, defined on G. As before, the open sets $h(G)$ and $-h(G)$ are disjoint, so if P is any point of $-h(G)$, then the holomorphic function

$$\tau(z) = \frac{1}{h(\sigma(z)) - P} \qquad (z \in U)$$

is bounded on U. By the the radial limit and boundary uniqueness theorems for bounded holomorphic functions ([Rdn '87, Theorem 11.28]), τ has the desired properties, hence so does $\sigma = (\frac{1}{\tau} + P)^2 + Q$.

(c) Let $\gamma(t) = \sigma^{-1}(\Gamma(t))$ for $0 \le t < \infty$, so γ is a curve in U. Because $\Gamma(t) \to \infty$, the univalence of σ forces $\gamma(t)$ to tend to ∂U as $t \to \infty$. Thus, the closure in the plane of the curve γ intersects ∂U in a continuum K, which we desire to show is a single point.

If this were not the case then K would be a non-trivial subarc of ∂U, hence for each $\omega \in K$, except possibly for the endpoints of K, the radius from 0 to ω would intersect γ infinitely often (see Figure 9.3 below).

Thus for each such ω there would be a sequence $t_n \to \infty$ (in general dependent on ω) such that

$$\gamma(t_n) = r_n\omega \to \omega,$$

while of course $\sigma(r_n\omega) = \Gamma(t_n) \to \infty$.

But according to part (a), σ has a radial limit at almost all points of ∂U, so the work of the last paragraph shows that this radial limit function would have to equal ∞ a.e. on K. But this contradicts part (b). Thus K must be a single point. $\square$

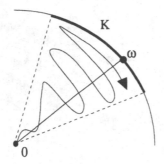

FIGURE 9.3. *The curve γ, the continuum K, and the radius to ω*

Remark. The proof of (b) above actually provides a much stronger result: σ^* is not constant on any subset of ∂U having positive measure. This follows from the correspondingly strengthened result for bounded analytic functions (see [Rdn '87, Theorem 17.18]). We do not, however, require this refinement.

Lemma 3 (Lindelöf's Theorem). *Suppose:*

(a) *f is a bounded holomorphic function on U,*

(b) *γ is a Jordan arc in U with an endpoint $\omega \in \partial U$, and*

(c) *$\lim f(z) = L \qquad (z \to \omega, \ z \in \gamma)$.*

Then $\lim_{r \to 1^-} f(r\omega) = L$.

Proof. We may without loss of generality assume that $\omega = +1$, $L = 0$, and that γ has one endpoint at the origin. Suppose first that γ has some last point r_0 of intersection with the interval $[0, 1)$. Let Ω be the Jordan subregion of U bounded by γ and its reflection γ_- in the real axis. By the Riemann Mapping Theorem and the Carathéodory Extension Theorem we can take the half of Ω bounded by γ and $[r_0, 1]$ onto the first quadrant, sending $+1$ to the origin, $[r_0, 1]$ to the imaginary axis with r_0 going to ∞, and γ to the positive real axis. By reflection this yields a univalent mapping τ taking Ω onto the upper half-plane Π^+, and extending to a homeomorphism of boundaries, as shown in Figure 9.4 below.
Upon defining $F = f \circ \tau^{-1}$, our task reduces to proving this:

Suppose F is holomorphic and bounded in Π^+ and continuous on $\overline{\Pi}^+ \backslash \{0\}$. If $F(x) \to 0$ as $x \to 0+$, then $F(iy) \to 0$ as $y \to 0+$.

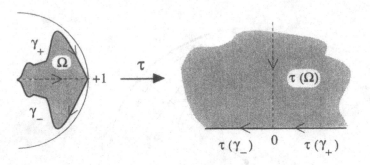

FIGURE 9.4. *Mapping Ω to the upper half-plane*

To prove this reduced version of Lindelöf's Theorem we may also assume without loss of generality that $|F| < 1$ on Π^+. Fix $0 < \varepsilon < 1$. Then there exists $a > 0$ such that $|F(x)| \le \varepsilon$ whenever $0 < x \le a$. For $z \in \Pi^+$ let

$$\theta(z) = \arg(z - a) - \arg z = \operatorname{Im} \log \frac{z - a}{z}.$$

where "arg" denotes the principal branch of the argument, and set $\omega(z) = \frac{1}{\pi}\theta(z)$. Then $\omega(z)$ is harmonic in the upper half-plane, and as summarized in Figure 9.5 below,

- $0 < \omega(z) < 1$ for all $z \in \Pi^+$,

- $\omega \equiv 1$ on the open interval $(0, a)$, and

- $\omega \equiv 0$ on $\mathbf{R} \backslash [0, a]$

(in other words, ω is the *harmonic measure* of the interval $(0, a)$).

It follows that the function

$$h(z) = \exp\left\{\frac{i}{\pi} \log \frac{z - a}{z}\right\} = \left\{\frac{z - a}{z}\right\}^{i/\pi}$$

is holomorphic on the upper half-plane, and has these properties:

- $|h(z)| = \exp\{-\omega(z)\} < 1$ for all $z \in \Pi^+$,

- $|h| \equiv e^{-1}$ on the open interval $(0, a)$, and

- $|h| \equiv 1$ on $\mathbf{R} \backslash [0, a]$.

Thus if $N = -\log \varepsilon$ we see that h^N is a bounded holomorphic function on Π^+ that is continuous on $\overline{\Pi}^+ \backslash \{0, a\}$, and

$$|F| \le |h^N| \qquad \text{on } \mathbf{R} \backslash \{0, a\}.$$

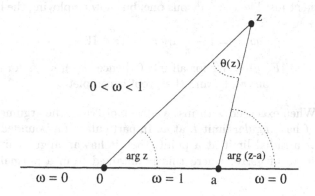

FIGURE 9.5. *Harmonic measure of* $(0, a)$

Thus, the Maximum Principle for bounded holomorphic functions insures that $|F| \leq |h^N|$ on all of Π^+. Specializing this to the imaginary axis we find that if $y > 0$ then

$$|F(iy)| \leq \exp\{-N\omega(iy)\}$$

$$= \exp\{-\frac{N}{\pi} \arctan \frac{a}{y}\}$$

$$\rightarrow e^{-N/2} = \sqrt{\varepsilon}$$

as $y \rightarrow 0+$, where the computation of $\omega(iy)$ in the second line is accomplished by redrawing Figure 9.5 with $z = iy$, and looking at the resulting right triangle. Thus

$$\limsup_{y \rightarrow 0+} |F(iy)| \leq \sqrt{\varepsilon}.$$

Since ε is any number strictly between 0 and 1 this shows that $F(iy) \rightarrow 0$ as $y \rightarrow 0+$.

Now suppose γ intersects $[0, 1)$ infinitely often. As before, we may assume that $|f| < 1$ on U, and by omitting a suitable initial arc of γ, say from 0 to r_0, we may further suppose that $|f| < \varepsilon$ on all of γ. We claim that $|f| < \sqrt{\varepsilon}$ on $[r_0, 1)$. Any parts of γ that are subintervals of the unit interval already obey this inequality. What remains is an at most countable collection γ_n of Jordan arcs, disjoint from the real axis except for endpoints $0 < a_n < b_n$ on the unit interval. For each such n we proceed as before to map the region Ω_n bounded by γ_n and its reflection in the real axis onto the upper half-plane, sending the interval (a_n, b_n) to the imaginary axis, but now sending γ_n to the *entire* positive real axis. The function f on U is now transformed into a holomorphic function F that is bounded by 1 on Π^+, and by ε on the positive real axis.

An argument just like the previous one, but now employing the harmonic function

$$\omega(z) = 1 - \frac{1}{\pi} \arg z \qquad (z \in \Pi^+),$$

establishes that $|F(iy)| < \sqrt{\varepsilon}$ for all $y > 0$, hence $|f(r)| < \sqrt{\varepsilon}$ for $a_n \leq r \leq b_n$. Thus $|f| < \sqrt{\varepsilon}$ on $[r_0, 1)$, and the proof is complete. $\qquad\qquad\square$

Remark. When executed with just a little more care, the argument above shows that f has *angular* limit L at ω. In particular, if a bounded analytic function has a radial limit at a point, then it has an angular limit there (cf. Exercise 13 of §9.10, where this is obtained from a normal families argument).

9.9   Proof of the No-Sectors Theorem

We begin with the crucial case $n = 1$ of the theorem, for which the situation is as follows:

- G is an unbounded, strictly starlike region, not the whole plane.

- σ is the univalent map that takes U onto G, with $\sigma(0) = 0$.

- λ is a fixed positive number, $0 < \lambda < 1$.

- φ is the univalent self-map of U defined by $\varphi(z) = \sigma^{-1}(\lambda\sigma(z))$.

Our goal is to show that under these hypotheses, if C_φ is not compact on H^2 then G contains a sector. In view of the angular derivative criterion of §4.2, the desired result can be rephrased as follows:

> If φ obeys the hypotheses above and has an angular derivative at some point of ∂U then G contains a sector.

Proof. We may assume that φ has an angular derivative at $+1$. In particular $\varphi(1) \overset{\text{def}}{=} \lim_{r \to 1-} \varphi(r)$ exists and belongs to ∂U.

STEP I. $\lim_{r \to 1-} \sigma(r) = \infty$.

Suppose this were not the case. Then, as in the proof of the Twisted Sector Theorem (§9.6), there would be a sequence $r_n \to 1-$ and a point $P \in \overline{G}$ such that $\sigma(r_n) \to P$. By the strict starlikeness of G we would have $\lambda P \in G$. Since φ, λ, and σ obey Schröder's equation,

$$\sigma(\varphi(r_n)) = \lambda\sigma(r_n) \to \lambda P \in G,$$

so by the univalence of σ,

$$\varphi(r_n) \to \sigma^{-1}(\lambda P) \in U,$$

which contradicts the fact that $\varphi(r_n) \to \varphi(1) \in \partial U$.

STEP II. Let $\theta(r)$ be a continuous determination of the argument of $\sigma(r)$. If it happens that $\lim_{r \to 1-} \theta(r)$ does not exist, then we are done! For in this case

$$\alpha_- \overset{\text{def}}{=} \liminf_{r \to 1-} \theta(r) < \limsup_{r \to 1-} \theta(r) \overset{\text{def}}{=} \alpha_+$$

and the strict starlikeness of G insures that for each $\alpha_- < \theta < \alpha_+$ the ray $\{te^{i\theta} : t \geq 0\}$ lies entirely in G (we invite the reader to draw a picture and supply the details). Thus, the entire sector $\{\alpha_- < \arg w < \alpha_+\}$ lies in G.

STEP III. The heart of the proof, then, is the case where the limit

$$\alpha \overset{\text{def}}{=} \lim_{r \to 1-} \theta(r)$$

exists.

In this step and the next we will show that the existence of this limit forces $+1$ to be a boundary fixed point of φ. Once this has been done then (recalling that $+1$ is also the point where the angular derivative of φ is assumed to exist) the Twisted Sector Theorem will guarantee that G contains a twisted sector about the curve $\sigma([0,1))$, after which a little extra argument using the structure provided by strict starlikeness will show that G contains a real sector.

Now, down to business. By Lemma 1 of the last section, the ray

$$L_\alpha = \{te^{i\alpha} : t \geq 0\}$$

lies in G. The proof from here on involves two pairs of Jordan curves:

- The line L_α and the curve $C \overset{\text{def}}{=} \sigma([0,1))$ in G, both of which join the origin to ∞, and

- the segment $[0,1)$ and the curve $\gamma = \sigma^{-1}(L_\alpha)$ in U.

By Lemma 2(c) of the last section, the curve γ, which begins at the origin, ends at some point $\omega \in \partial U$.

We claim that ω is a boundary fixed point of φ.

Note first that $\varphi(\gamma) = \gamma$. Indeed, since $\lambda > 0$ we have $\lambda L_\alpha = L_\alpha$, hence by Schröder's equation,

$$\sigma(\varphi(\gamma)) = \lambda\sigma(\gamma) = \lambda L_\alpha = L_\alpha,$$

whereupon

$$\varphi(\gamma) = \sigma^{-1}(L_\alpha) = \gamma.$$

Thus $\varphi(z) \to \omega$ as $z \to \omega$ through γ, so Lindelöf's theorem (Lemma 3 of the last section) yields

$$\varphi(\omega) \overset{\text{def}}{=} \lim_{r \to 1-} \varphi(r\omega) = \omega,$$

which is the desired result.

STEP IV. So far we know these facts:

(a) In G, the ray L_a and the Jordan curve $C = \sigma([0,1))$ both join the origin to ∞, and C is "asymptotically parallel" to L_α in the sense that $\arg w \to \alpha$ as $w \to \infty$ through C.

(b) In U, the Jordan curve $\gamma = \sigma^{-1}(L_\alpha)$ joins the origin to a point $\omega \in \partial U$, and ω is a boundary fixed point of φ.

(c) By hypothesis, φ has an angular derivative at $+1$.

We *claim that* $\omega = +1$.

If γ intersects the segment $[0,1)$ at a sequence of points approaching $+1$, then this is clear from (b) above, so let us assume that there is some final point of intersection $p \in [0,1)$. Redefine γ to be the part of $\sigma^{-1}(L_\alpha)$ that joins p to ω, and C to be the image of the new γ under σ. Let V denote the simply connected subregion of U bounded by the segment $[p,1]$, the curve γ, and allowing for the possibility that $\omega \neq 1$ (which, remember, we want to show does not happen), the arc from ω to $+1$. Over in G, let W be the simply connected region that lies "between" C and L_α, as shown in Figure 9.6 below.

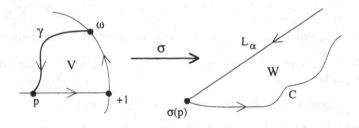

FIGURE 9.6. *The regions V and W*

Then $\sigma(V) \subset W$ because conformality preserves orientation. To show that $\omega = +1$ it suffices to prove that $\sigma(V) = W$, for once this is done Carathéodory's Theorem (applied to Jordan subregions of the Riemann Sphere) will insure that σ extends to a homeomorphism of ∂V onto ∂W (with ∞ regarded as a point of ∂W). Since $\sigma(\omega) = \sigma(1) = \infty$, this will imply that $\omega = 1$ as desired.

The first step in proving that $\sigma(V) = W$ is to argue that $\sigma(V) = W \cap G$. Since $\sigma(V)$ clearly lies in both W and G, we need only prove the containment "$\supset$." Suppose $Q \in W \cap G$. Then membership in G provides a point $q \in U$ with $\sigma(q) = Q$. Now the unit disc falls into three disjoint connected sets: the Jordan curve $\gamma_1 = \gamma \cup [p,1)$, the open set V, and the remaining open set $V_1 = U \backslash \overline{V}$. Since σ is a univalent function that takes γ_1 onto the union of C and L_α in G, and takes V into W, it must therefore take V_1 into $C \backslash \overline{W}$. It follows that the point q cannot lie in either the curve γ_1 or the open set V_1, hence it must lie in V. Thus $Q = \sigma(q) \in \sigma(V)$, which proves the desired inclusion.

Now suppose, for the sake of contradiction, that $W \neq \sigma(V)$, so some point $Q \in W$ does not lie in $\sigma(V)$. Since $\sigma(V) = W \cap G$, this can only happen if $Q \notin G$. By strict starlikeness, G must be disjoint from the entire half-line

$$L(Q) = \{tQ : t \geq 1\}.$$

Recall that L_α and C are "asymptotically parallel," so C must intersect $L(Q)$, as shown in Figure 9.7 below. But this contradicts the fact that $C \subset G$. Thus $\sigma(V) = W$, as desired.

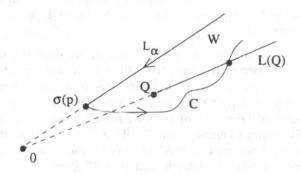

FIGURE 9.7. *The (nonexistent) line* $L(Q)$

STEP V. At this point we have established that, under the hypotheses of Step III, the boundary fixed point ω of φ is $+1$, the point where φ is assumed to have an angular derivative. Therefore G contains a twisted sector about $C = \sigma([0, 1))$. Recall that we are assuming as part of the hypotheses of Step III that

$$\lim_{r \to 1-} \theta(r) = \alpha, \tag{11}$$

where $\theta(r)$ is a continuous determination of the argument of $\sigma(r)$ for $0 \leq r < 1$. Let $P(r)$ denote the orthogonal projection of $\sigma(r)$ onto L_α. Our twisted sector conclusion asserts that there exists a positive number ε such that

$$\delta_G(\sigma(r)) \geq \varepsilon |\sigma(r)| \geq \varepsilon |P(r)| \tag{12}$$

for all $0 \leq r < 1$.

Because of (11) we may choose $0 < R < 1$ so that

$$|\tan(\theta(r) - \alpha)| < \frac{\varepsilon}{2}$$

whenever $R < r < 1$. For such r a little trigonometry shows that

$$|\sigma(r) - P(r)| = |P(r)||\tan(\theta(r) - \alpha)| < |P(r)|\frac{\varepsilon}{2},$$

and this, along with (12) above and the triangle inequality shows that

$$\begin{aligned} \delta_G(P(r)) \ &\geq\ \delta_G(\sigma(r)) - |P(r) - \sigma(r)| \\[2mm] &>\ \varepsilon|P(r)| - \frac{\varepsilon}{2}|P(r)| \\[2mm] &=\ \frac{\varepsilon}{2}|P(r)|. \end{aligned}$$

Since $P(r) \to \infty$ as $r \to 1-$, this estimate shows that G contains a sector about L_α, and completes the proof of the "$n = 1$" case of the No-Sectors Theorem.

FINAL STEP. To prove the rest of the Theorem, we drop the assumption that $\lambda > 0$. It is enough to prove that $\lambda^n > 0$ for some positive integer n. For then the previous case, with λ^n in place of λ will give the desired result: if C_φ^n is not compact then G contains a sector.

The Corollary of Lemma 1 of the last section asserts that since G is strictly starlike it must contain a ray L. Since G is taken into itself under multiplication by λ we must have $\lambda^n \cdot L \subset G$ for each positive integer n.

Suppose for the sake of contradiction that λ^n is never positive. Then the argument of λ is an irrational multiple of π, so the set of unimodular complex numbers

$$\{\lambda^n/|\lambda^n| : n = 1, 2, 3, \dots\}$$

is dense in the unit circle. It follows that the set

$$\bigcup_{n=1}^{\infty} \lambda^n \cdot L$$

is dense in $\mathbf{C}$, hence by Lemma 1 again, $G = \mathbf{C}$. But we are assuming that $G \neq \mathbf{C}$, so there must be some positive integer n for which $\lambda^n > 0$. This completes the proof of the No-Sectors Theorem. $\qquad\square$

9.10   Exercises

1. Show that if G is the right half-plane, then

$$h_G(w) = \frac{1}{\operatorname{Re} w} \qquad (w \in G).$$

2. For $0 < R < \infty$ and $p \in U$ let $C_p(R)$ denote the hyperbolic circle of radius R and center p. That is,

$$C_p(R) = \{z \in U : \rho_U(p, z) = R\}.$$

Show that $\ell_U(C_p(R)) = 2\pi \sinh R$.

3. Show that for any simply connected domain G, not the whole plane, ρ_G is a complete metric on G that induces the Euclidean topology.

4. Show that for each of the non-tangential lenses L_α $(0 < \alpha < 1)$ introduced in §2.3,

$$\partial L_\alpha = \{z \in U : \rho_U(z, \text{real axis}) \equiv \text{const.}\},$$

where the constant depends on α.

5. Suppose $0 < r < 1$. Show that for every $p, q \in U$,

$$\rho_U(rp, rq) \le r\rho_U(p, q).$$

Suggestion: First derive the corresponding inequality for the hyperbolic lengths of curves.

6. Show that every holomorphic self-map φ of U is a *contraction* in the hyperbolic metric, that is:

$$\rho_U(\varphi(p), \varphi(q)) \le \rho_U(p, q)$$

for all $p, q \in U$.

7. Use the hyperbolic metric to prove this special case of the Denjoy-Wolff Theorem of Chapter 5:

> If φ is a holomorphic self-map of U with $\|\varphi\|_\infty < 1$, then φ has a fixed point in U.

Suggestion: Use Exercises 5 and 6 above to show that

$$\rho_U(\varphi(p), \varphi(q)) \le \|\varphi\|_\infty \rho_U(p, q).$$

Then use the hyperbolic completeness of U and the Banach Contraction Mapping Principle (see e.g. [Frd '82, Th. 3.8.2, page 119]).

8. More generally, suppose G is a simply connected domain contained in a non-tangential lens L_α for some $0 < \alpha < 1$. Show that if φ maps U into G, then

$$\rho_U(\varphi(z_1), \varphi(z_2)) \leq \alpha \rho_U(z_1, z_2)$$

for every pair of points $z_1, z_2 \in U$. Conclude that φ must have a fixed point in U.

Suggestion: Show that it suffices to prove the result for φ a conformal mapping of U onto L_α. For this it is easier to transfer the problem into the right half-plane, where the corresponding mapping is simply the principal branch of the α-th power function. Then use Exercise 1 above to get started on the estimate.

Remark: We have known since §2.3 that φ induces a compact composition operator on H^2, so the existence of an interior fixed point follows from the Corollary in §5.5. The point of this problem is to derive the result using only hyperbolic geometry and the standard contraction mapping theorem.

9. Show that, up to constant multiples, h_U is the unique positive continuous distortion function on U that induces a conformally invariant metric.

10. *Hyperbolic area.* Define the *hyperbolic area* of a Borel subset E of U to be:

$$\nu(E) = \int_E \frac{dx\,dy}{(1 - |z|^2)^2}.$$

  (a) Show that the measure ν is conformally invariant in the sense that $\nu(\alpha(E)) = \nu(E)$ for each $\alpha \in \mathrm{Aut}(U)$.

  (b) Compute the hyperbolic area of a hyperbolic disc of radius R (the interior of the circle $C_p(R)$ of Exercise 2 above).

11. Derive the upper estimate

$$h_G(w) \leq \frac{2}{\delta_G(w)}$$

mentioned in the remark following Corollary 2 of §9.5.

Suggestion: Use the Schwarz Lemma to prove that if f is univalent on U, $f(0) = 0$ and $f'(0) = 1$, then $f(U)$ does *not* contain a disc centered at the origin of radius > 1 (i.e., if $G = f(U)$ then $\delta_G(0) \leq 1$). Then proceed as in the proof of Corollary 2.

12. *Bloch Functions.* Suppose $f \in H(U)$ and $G = f(U)$. An open disc $\Delta \subset G$ is said to be *schlicht* if $\Delta = f(V)$, where V is an open subset of U on which f is univalent. For $w \in G$ let $r(w)$ denote the radius of the largest schlicht disc in G that is centered at w.

   (a) Show that
   $$r(f(z)) \leq |f'(z)|(1 - |z|^2)$$
   for every $z \in U$.

   Suggestion: First prove the result for $z = 0$ by an appropriate application of the Schwarz Lemma. Then reduce the general case to this one by a conformal automorphism of U.

   (b) *Bloch's Theorem* asserts that there is a positive constant β such that if $f \in H(U)$ and $|f'(0)| = 1$, then $r(f(0)) \geq \beta$ (the exact value of β is still not known). Use Bloch's Theorem to show that
   $$\beta|f'(z)|(1 - |z|^2) \leq r(f(z))$$
   for each $z \in U$.

   (c) The *Bloch space* $\mathcal{B}$ consists of all $f \in H(U)$ for which
   $$\sup_{z \in U} |f'(z)|(1 - |z|^2) < \infty.$$
   Show that $f \in \mathcal{B}$ if and only if $f(U)$ does not contain arbitrarily large schlicht discs.

13. Using normal families, prove that if F is a bounded holomorphic function on the upper half-plane, and $f(iy) \to L$ as $y \to 0+$, then $\angle \lim_{w \to 0} F(w) = L$.

   Suggestion: Consider the normal family $\{F(2^{-n}w)\}_0^\infty$, and look at what happens to a convergent subsequence in the part of a sector that lies between lines $y = 1$ and $y = 1/2$.

14. This exercise shows that for each of the examples of §9.3, $\varphi(U)$ is contained in a lens, hence the compactness of the induced operators can also be deduced from the results of Chapter 2. It also shows that $\varphi(U)$ can lie in a lens even though the Königs function of φ contains arbitrarily large discs.

   Suppose G is a domain as shown below, where the boundary has the form $|y| = f(x)$ for $x \geq -a$ (some $a > 0$), and the function f is non-negative, differentiable, and increasing on $[-a, \infty)$, with derivative decreasing to zero. Let φ be the univalent self-map of U whose Königs model is $(M_{1/2}, G)$. Suppose that in addition, for some $A > 1/2$ we have $f(x) \geq Af(2x)$ for all x. (Examples: For $0 < \alpha < 1$ let $f(x) = (x + 1)^\alpha$ for $x > -1$.)

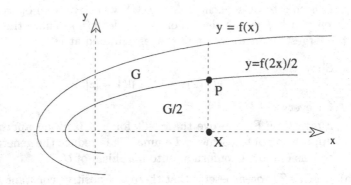

FIGURE 9.8. *The regions G and $\frac{1}{2}G$*

Show that $\varphi(U)$ lies in a lens shaped region L_α.

Suggestion: According to Exercise 4 above, it suffices to show that the hyperbolic distance from each boundary point of $\frac{1}{2}G$ to the real axis is bounded by a fixed constant. Let $P = (X, Y) \in \frac{1}{2}G$, and let Γ be the vertical segment from $(X, 0)$ to P. Then verify that

$$\rho_G(P, \text{real axis}) \leq \ell_G(\Gamma) \leq 2 \int_\Gamma \frac{|dw|}{\delta_G(w)} \leq \frac{f(2X)}{\delta_G(P)}$$

where the second inequality comes from Exercise 11 above. Observing that the boundary of $\frac{1}{2}G$ has equation $|y| = \frac{1}{2}f(2x)$, show that

$$\delta_G(P) \approx f(X) - \frac{1}{2}f(2X)$$

as $X \to \infty$, from which the desired result follows.

9.11   Notes

The major results and examples in this chapter come from [SSS '92]. In that paper it is further shown that without any assumptions of starlikeness:

If G contains a twisted sector, then $\sigma \notin H^p$ for some $p < \infty$.

This result is then used along with the Twisted Sector Theorem to prove [SSS '92, §1, Corollary 2]:

> If G has only finitely many hyperbolic geodesics to ∞ then the
> following are equivalent:
>
> (a) Some power of C_φ is compact.
> (b) $\sigma \in H^p$ for all $p < \infty$.
> (c) G contains no twisted sector.

Roughly speaking, each geodesic to ∞ corresponds to a "tube" in G that
connects the origin to ∞. Examples given in [SSS '92] show that the result
can fail if G has infinitely many geodesics to ∞, and no extra assumption
like strict starlikeness is made.

Lindelöf's Theorem (Lemma 3 of §9.8) occurs in [Lnd '15]. The proof
given here is essentially the one found in §309 of [Cth '54]. The Distance
Lemma of §9.5 is Lemma 2.1 of [GgP '76]. Exercise 14 above is due to
Wayne Smith.

In this chapter we developed just enough hyperbolic geometry to suit
our purposes. For more on this fascinating subject, including extensions to
more general domains in the plane, as well as in higher real and complex
dimensions, see [Brd '82, Chapter 7], [Krz '90], and [Krz '92].

10

Compactness: General Case

So far our pursuit of compactness has dwelt on geometric issues associated with univalent inducing maps (cf. Chapters 3, 4, and 9). In this chapter we abandon univalence, and attack the compactness problem for composition operators induced by arbitrary holomorphic self-maps of the disc. As you might guess, the solution involves not only the geometry of the inducing map's image, but its "affinity" for each of the points in this image.

10.1 Motivation

Our approach to the compactness problem for general inducing maps will feature many of the same ideas we used to solve the univalent problem (Chapter 3). Recall the first of these: In §3.1 we derived an area-integral estimate for the H^2-norm that had the form

$$\|f\|^2 \approx |f(0)|^2 + \int_U |f'(z)|^2(1 - |z|)\, dA(z) \qquad (1)$$

with dA denoting Lebesgue area measure on U, normalized to have total mass one (i.e., $dA(z) = dx\,dy/\pi$). For φ univalent, the change-of-variable formula for multiple integrals resulted in a norm estimate for $C_\varphi f = f \circ \varphi$ that looked like this:

$$\|C_\varphi f\|^2 \approx |f(\varphi(0))|^2 + \int_{\varphi(U)} |f'(w)|^2 \Omega(w)(1 - |w|)\, dA(w) \qquad (2)$$

where

$$\Omega(\varphi(z)) = \frac{1 - |z|}{1 - |\varphi(z)|} \qquad (z \in U).$$

At this point the Schwarz Lemma provided the crucial element of the proof:

$$\Omega(\varphi(z)) = O(1) \quad \text{as} \quad |z| \to 1 - .$$

For φ univalent, this inequality, along with the norm expressions above, "explained" the boundedness of C_φ on H^2, and led us to establish the "Univalent Compactness Theorem:"

$$C_\varphi \text{ is compact on } H^2 \quad \Longleftrightarrow \quad \Omega(\varphi(z)) = o(1) \text{ as } |z| \to 1-.$$

In Chapter 4 the condition on the right, rewritten as

$$\lim_{|z| \to 1-} \frac{1 - |\varphi(z)|}{1 - |z|} = \infty,$$

was shown to be equivalent to the "Angular Derivative Criterion":

φ has an angular derivative at no point of ∂U.

We noted in Chapter 3 that the Angular Derivative Criterion remains necessary for compactness, even in the absence of univalence. In the next section we will see that without some extra valence restriction, the criterion fails, in general, to be *sufficient* for compactness. In the remaining sections we modify the univalent solution according to the strategy outlined below to characterize compactness for arbitrary maps φ.

Instead of using the approximate representation (1) above for the H^2 norm, it is more convenient put a little more care into the proof, and arrive at a precise formula (see Chapter 3, Exercise 1):

The Littlewood-Paley Identity. *If f is holomorphic on U, then*

$$\|f\|^2 = |f(0)|^2 + 2 \int_U |f'(z)|^2 \log \frac{1}{|z|} \, dA(z) \tag{3}$$

where, as always, "$\| \ \|$" denotes the H^2 norm, and "$\|f\| = \infty$" means "$f \notin H^2$."

The next step is to determine the effect of the change of variable $w = \varphi(z)$ in the Littlewood-Paley integral. The result turns out to be the expression you would get if φ were univalent, but now summed over all possible preimages. More precisely:

The change-of-variable formula. *Suppose φ is holomorphic on U. Then*

$$\|C_\varphi f\|^2 = |f(\varphi(0))|^2 + 2 \int_U |f'(w)|^2 N_\varphi(w)\, dA(w) \qquad (4)$$

where

$$N_\varphi(w) = \begin{cases} \displaystyle\sum_{z \in \varphi^{-1}\{w\}} \log \frac{1}{|z|} & \text{if } w \in \varphi(U) \\[2em] 0 & \text{if } w \notin \varphi(U). \end{cases} \qquad (5)$$

Discussion. N_φ is the *Nevanlinna Counting Function* of φ. In its definition, the inverse image $\varphi^{-1}\{w\}$ denotes *zero-sequence* of the function $\varphi - w$, as defined in our discussion of Jensen's formula in §7.3. So $\varphi^{-1}\{w\}$ is a sequence whose terms are the points of the set-theoretic inverse image, written in order of increasing moduli, with each point occurring as many times as its multiplicity.

Having said this, we hasten to add that multiplicity arises only for values w whose pre-image contains points at which φ' vanishes. Since the offending values and their pre-images form an at most countable set, it will be possible to ignore them in the area-integral computations involved in proving the change-of-variable formula (4).

For $w \neq \varphi(0)$, the convergence of the sum that defines the counting function is not in doubt. Indeed, the function $\varphi - w$ is bounded, so by the Zero-sequence Theorem of §7.3,

$$\sum_{z \in \varphi^{-1}\{w\}} (1 - |z|) < \infty.$$

Since $1 - |z| \approx \log \frac{1}{|z|}$ for $z \in U$ bounded away from the origin, the convergence of the counting-function sum is actually equivalent to the above summability condition.

The number $N_\varphi(w)$ provides a measure of the "affinity" that φ has for the value w. In the computation of $N_\varphi(w)$, the counting function weights each pre-image by the product of its multiplicity and a factor that is (away from the origin) essentially its distance from ∂U. So pre-images that lie deeper inside the unit disc count more. At the extremes, $N_\varphi(w) = 0$ if w is not a value of φ, and $N_\varphi(w) = \infty$ if $w = \varphi(0)$ (which explains our preference, in what follows, for avoiding this value of w). In all other cases, $0 < N_\varphi(w) < \infty$.

For another interpretation of the counting function as an "averaged multiplicity function," see Exercise 3 below.

Remark. Note that in case φ is univalent, the sum that defines N_φ has just one term, in which case (4) reduces to (2), with the integrable singularity of $\log(1/|w|)$ at the origin causing no difficulty because the functions being integrated are well-behaved there.

With the change-of-variable formula in hand, the final motivation for our solution comes from a striking generalization of the Schwarz Lemma (due appropriately enough to Littlewood), which asserts that for any holomorphic self-map φ of U,

$$N_\varphi(w) = O\left(\log \frac{1}{|w|}\right) \quad \text{as} \quad |w| \to 1-. \tag{6}$$

Thus (6), together with the Littlewood-Paley Identity and the change-of-variable formula, shows that the "universal" boundedness of composition operators on H^2 arises from the universal estimate (6) on the decay of counting functions for holomorphic self-maps of the disc. More importantly, this estimate, along with the "big oh–little oh" heuristics of Chapter 3 leads us to conjecture the solution of the general compactness problem.

The Compactness Theorem. *Suppose φ is a holomorphic self-map of U. Then C_φ is compact on H^2 if and only if*

$$\lim_{|w|\to 1-} \frac{N_\varphi(w)}{\log \frac{1}{|w|}} = 0. \tag{7}$$

Of course we could as well write (6) as: $N_\varphi(w) = O(1 - |w|)$, and state the compactness criterion as: $N_\varphi(w) = o(1 - |w|)$. Our preference for the logarithmic version of the distance to the boundary originates with the definition of the counting function in terms of the "logarithmic distance" of preimages to the boundary. For reasons that trace back to Jensen's Formula (§7.3), this logarithmic distance, rather than the usual one, is the quantity most closely associated with the relevant function theory.

Note that the Compactness Theorem reduces to the result of Chapter 3 if φ is univalent. Once (6) is proved, the proof that (7) is *sufficient* for compactness will be similar to the corresponding univalent argument. However the proof of its *necessity* will be more subtle, relying not on operator theory, but on a "subharmonic" property of the counting function. We will further discover that behind both (6) and this subharmonic property lurks Jensen's formula of §7.3.

10.2   Inadequacy of Angular Derivatives

In this section we give an example of a holomorphic self-map of U that has an angular derivative at *no* point of ∂U, but which nevertheless induces a non-compact composition operator on H^2. Thus the non-existence of the angular derivative, which characterizes compactness in the univalent case, does not furnish a sufficient condition for general inducing maps.

Blaschke Products. Suppose $\{b_n\}_1^\infty$ is a sequence of (not necessarily distinct) points of U that tends to the boundary rapidly enough so that

$$\sum_{n=1}^\infty (1 - |b_n|) < \infty. \tag{8}$$

Condition (8) has surfaced previously as a restriction on the zero-sequences of non-trivial H^2 functions, and again in our discussion of the Counting Function. The work that follows will show that it actually *characterizes* the zero-sequences of H^2 functions, and of bounded holomorphic functions as well!

Recall that every H^2 function, and in particular every bounded holomorphic function f, has a radial limit

$$f^*(\zeta) = \lim_{r \to 1-} f(r\zeta)$$

at almost every point $\zeta \in \partial U$.

Proposition. *Suppose $\{b_n\}$ is a sequence in $U \setminus \{0\}$ that satisfies condition (8) above. Then the infinite product*

$$\prod_{n=1}^\infty \frac{|b_n|}{b_n} \frac{b_n - z}{1 - \overline{b_n} z}$$

converges uniformly on compact subsets of U to a holomorphic function B with these properties:

(a) *B vanishes precisely at the points $\{b_n\}$, and with the correct multiplicities (i.e., $\{b_n\}$ is the zero-sequence of B),*

(b) *$|B(z)| < 1$ for every $z \in U$,*

(c) *$|B^*| = 1$ a.e. on ∂U.*

The holomorphic function B is called the *Blaschke product* with zero-sequence $\{b_n\}$, and (8) is called the *Blaschke condition* on $\{b_n\}$.

Remark. The unimodular weights $\{|b_n|/b_n\}$ are chosen so that $B(0) = \prod_n |b_n|$. Thus, the initial conclusion of the Proposition asserts that the infinite product converges uniformly on compact subsets of U if and only if it converges at the origin.

Proof of Proposition. The n-th term of the infinite product is a unimodular multiple of the special automorphism α_{b_n}. Thus part (b) of the Proposition will follow from the convergence asserted for the product, and for this convergence it suffices to show that the sum

$$\sum_{n=1}^{\infty} \left| 1 - \frac{|b_n|}{b_n} \alpha_{b_n} \right| \tag{9}$$

converges uniformly on compact subsets of U (cf. [Rdn '87, §15.1–15.6]). One checks easily that for $z \in U$,

$$\left| 1 - \frac{|b_n|}{b_n} \alpha_{b_n}(z) \right| = \left| \frac{1 + |b_n| z / b_n}{1 - \overline{b}_n z} \right| (1 - |b_n|)$$

$$\leq \frac{1 + |z|}{1 - |z|} (1 - |b_n|),$$

and this, along with the Blaschke condition (8) gives the desired convergence for the sum (9). This proves that B is holomorphic in U, and that $|B| \leq 1$ on U. Since B has zeros in U, but is not zero at the origin, it is not identically zero, so the Maximum Principle implies that $|B| < 1$ in U. Thus (b) is established, and (a) now follows from Hurwitz's Theorem ([Rdn '87] Ch. 10, Problem 20).

For assertion (c), let N be a positive integer, and write

$$B_N(z) = \prod_{n=N}^{\infty} \frac{|b_n|}{b_n} \alpha_{b_n}(z),$$

the N-th remainder of the infinite product for B. Fix $0 < r < 1$. Then for all sufficiently large positive integers N we have $|b_n| > r$ if $n \geq N$ (we are assuming that the sequence $\{b_n\}$ is infinite; otherwise there is nothing to prove). Thus B_N does not vanish in some neighborhood of the closed disc $r\overline{U}$, hence $\log |B_N|$ is harmonic there. By the Mean Value Property of harmonic functions ([Rdn '87], §11.12–13),

$$\log \prod_{n=N}^{\infty} |b_n| = \log |B_N(0)| = \frac{1}{2\pi} \int_{-\pi}^{\pi} \log |B_N(re^{i\theta})| \, d\theta. \tag{10}$$

Now the Arithmetic-Geometric Mean inequality asserts that if μ is a probability measure (positive, with total mass one) and $g \in L^1(\mu)$, then

$$\exp \left\{ \int g \, d\mu \right\} \leq \int e^g \, d\mu$$

([Rdn '87], §3.3, inequality (4)). Upon applying this to the unit circle, with

$$g(e^{i\theta}) = \log |B_N(re^{i\theta})| = \frac{1}{2} \log \{ |B_N(re^{i\theta})|^2 \},$$

we obtain

$$\exp\left\{\frac{1}{2\pi}\int_{-\pi}^{\pi}\log|B_N(re^{i\theta})|\,d\theta\right\}\leq\left\{\frac{1}{2\pi}\int_{-\pi}^{\pi}|B_N(re^{i\theta})|^2\,d\theta\right\}^{1/2}.\qquad(11)$$

Furthermore we know from §1.2 that as $r \to 1-$,

$$\frac{1}{2\pi}\int_{-\pi}^{\pi}|B_N(re^{i\theta})|^2\,d\theta \nearrow \|B_N\|^2 = \|B\|^2,\qquad(12)$$

where the equality follows from the representation of the H^2-norm as a boundary integral (§2.3), and the fact that $|B_N^*| \equiv |B^*|$ on ∂U.

From (10) through (12) follows

$$\prod_{n=N}^{\infty}|b_n| \leq \|B\|,$$

and since the product $\prod_n |b_n|$ converges, the left-hand side of this inequality tends to 1 as $N \to \infty$. Thus $1 \leq \|B\|$. On the other hand, we know that $|B| \leq 1$ on U, so $|B^*| \leq 1$ a.e. on ∂U. Thus by the boundary-integral representation of the norm (§2.3),

$$1 \leq \|B\|^2 = \frac{1}{2\pi}\int_{-\pi}^{\pi}|B^*(e^{i\theta})|^2 d\theta \leq 1$$

so there is equality throughout. In particular, $|B^*| = 1$ a.e. on ∂U. □

Recall from §2.5 that if φ is a holomorphic self-map of U with $|\varphi^*| = 1$ on a set of positive measure in ∂U, then C_φ is not compact on H^2. In particular:

> *Every Blaschke product induces a non-compact composition operator on H^2.*

Thus if we can arrange the zeros of a Blaschke product so as to insure that the angular derivative exists nowhere on ∂U, then we will have constructed an example that shows the angular derivative criterion is not, in general, sufficient for compactness. The result that makes this possible is:

Frostman's Theorem. *Let B be a Blaschke product with zero-sequence $\{b_n\}$, and ζ a point of ∂U. Then B has an angular derivative at ζ if and only if*

$$\sum_{n=1}^{\infty}\frac{1-|b_n|}{|\zeta - b_n|^2} < \infty.\qquad(13)$$

Proof. Only the necessity of (13) is required for our purposes, so we prove this and leave the sufficiency to the reader (Exercise 1 below). We assume that the sequence $\{b_n\}$ is infinite—otherwise there is nothing to prove—and we may of course take $\zeta = +1$. Thus we are assuming that B has an angular derivative $B'(1)$ at $+1$, and we wish to show that (13) holds. To this end let

$$B_N = \prod_{n=1}^{N} \frac{|b_n|}{b_n} \alpha_{b_n},$$

the N-th partial product of the infinite product for B (*not* the N-th remainder, as occurred in the previous argument). Since each term of the product is < 1 in magnitude, we have $|B| < |B_N|$ at each point of U, hence

$$\frac{1 - |B_N(r)|}{1 - r} < \frac{1 - |B(r)|}{1 - r} \tag{14}$$

for each $0 < r \leq 1$. Therefore the existence of the angular derivative $B'(1)$ gives rise to these estimates:

$$
\begin{aligned}
|B'(1)| &= \lim_{r \to 1-} \left| \frac{B(1) - B(r)}{1 - r} \right| \\[2mm]
&\geq \limsup_{r \to 1-} \frac{1 - |B(r)|}{1 - r} \\[2mm]
&\geq \limsup_{r \to 1-} \frac{1 - |B_N(r)|}{1 - r} \\[2mm]
&= |B_N'(1)|,
\end{aligned}
$$

where the second line follows from the fact that $|B(1)| = 1$, the third one from (14) above, and the last one from the Julia-Carathéodory Theorem. Note that in this last line the *existence* of $B_N'(1)$ is not in doubt; its *magnitude* is what we require, and for this the Julia-Carathéodory Theorem is required to get the proper estimate.

Now logarithmic differentiation yields

$$
\begin{aligned}
\frac{B_N'(1)}{B_N(1)} &= \sum_{n=1}^{N} \frac{\alpha_{b_n}'(1)}{\alpha_{b_n}(1)} \\[2mm]
&= \sum_{n=1}^{N} \frac{1 - |b_n|^2}{(1 - \overline{b_n})^2} \cdot \frac{1 - \overline{b_n}}{1 - b_n} \\[2mm]
&= \sum_{n=1}^{N} \frac{1 - |b_n|^2}{(1 - \overline{b_n})(1 - b_n)}.
\end{aligned}
$$

Since $|B_N(1)| = 1$ this calculation, along with the previous estimate, yields

$$\sum_{n=1}^{N} \frac{1 - |b_n|^2}{|1 - b_n|^2} = |B_N'(1)| \leq |B'(1)|,$$

for every positive integer N, hence

$$\sum_{n=1}^{\infty} \frac{1 - |b_n|^2}{|1 - b_n|^2} \leq |B'(1)| < \infty,$$

which, because $(1 - |b_n|) \leq 1 - |b_n|^2$, yields the desired result.   □

Remark. The argument above can be generalized to show that if φ is any holomorphic self-map of U that has an angular derivative at $\zeta \in \partial U$, then the zero-sequence of φ satisfies (13) (see Exercise 2 below).

Example. *There is a Blaschke product that has an angular derivative at no point of ∂U.*

Proof. By Frostman's Theorem, we need only arrange the prospective zeros $\{b_n\}$ so that that the Blaschke condition $\sum(1 - |b_n|) < \infty$ is satisfied, yet

$$\sum \frac{1 - |b_n|}{|\zeta - b_n|^2} = \infty \quad \text{for each } \zeta \in \partial U.$$

To this end let $\{I_n\}_1^{\infty}$ be a sequence of contiguous subarcs of ∂U, with

$$\text{length of } I_n = \frac{1}{n}.$$

Since this sequence of arcs has infinite total length, it wraps infinitely often around the unit circle, so each point of ∂U belongs to infinitely many I_n.

For each positive integer n let ζ_n be the center of I_n, and set $b_n = (1 - n^{-2})\zeta_n$. Then $1 - |b_n| = n^{-2}$, so the sequence $\{b_n\}$ satisfies the Blaschke condition, while a quick sketch shows that for each $\zeta \in I_n$,

$$|\zeta - b_n| < \frac{1}{2n} + \frac{1}{n^2} < \frac{2}{n}. \tag{15}$$

Now fix $\zeta \in \partial U$. Then $\zeta \in I_n$ for infinitely many indices n, for each of which we have the estimate (15) above. For each of these indices,

$$\frac{1 - |b_n|}{|\zeta - b_n|^2} \geq \frac{(1/n)^2}{(2/n)^2} = 1/4,$$

hence

$$\sum_{n=1}^{\infty} \frac{1 - |b_n|}{|\zeta - b_n|^2} = \infty,$$

as desired.   □

The Blaschke product constructed above furnishes the example we seek: a holomorphic self-map of U that has an angular derivative nowhere on ∂U, yet induces a non-compact composition operator on H^2.

10.3   Non-Univalent Changes of Variable

Having established that the solution to the univalent compactness problem does not work for the general case, we embark on the program outlined in §10.1 for solving the compactness problem once and for all. The first order of business is to establish the change-of-variable formula (4). This follows immediately from the Littlewood-Paley Identity (3), and the case $g = |f'|^2$ of the result below.

Proposition. *If g is a non-negative measurable function on U and φ a holomorphic self-map of U, then*

$$\int_U g(\varphi(z))|\varphi'(z)|^2 \log \frac{1}{|z|} \, dA(z) = \int_U g N_\varphi \, dA.$$

Proof. The derivative φ' vanishes on an at most countable subset Z of U with no limit point of U. At every point of $U \backslash Z$ there is an open set on which φ is a homeomorphism. Thus $U \backslash Z$ can be decomposed into an at most countable disjoint collection $\{R_n\}$ of "semi-closed" polar rectangles, on each of which φ is univalent. Let ψ_n be the inverse of the restriction of φ to R_n. Then the usual change-of-variable formula applied to the substitution $z = \psi_n(w)$ yields

$$\int_{R_n} g(\varphi(z))|\varphi'(z)|^2 \log \frac{1}{|z|} \, dA(z) = \int_U g\chi_n \log \frac{1}{|\psi_n|} \, dA$$

where χ_n is the characteristic function of $\varphi(R_n)$. Summing both sides on n, we obtain

$$\int_U g(\varphi(z))|\varphi'(z)|^2 \log \frac{1}{|z|} \, dA(z) = \int_U g \left\{ \sum_n \chi_n \log \frac{1}{|\psi_n|} \right\} \, dA.$$

Now for $w \in \varphi(U) \backslash \varphi(Z)$ the points of the inverse image $\varphi^{-1}\{w\}$ all have multiplicity one, so the term in braces on the right-hand side of the last displayed equation coincides a.e. on $\varphi(U)$ with $N_\varphi(w)$. The same is true for $w \notin \varphi(U)$, in which case both the term in braces and the counting function take the value zero. $\qquad\square$

10.4   Decay of the Counting Function

In this section we prove estimate (6), which shows that as you approach the unit circle, the Nevanlinna Counting Function decays to zero "like the distance to the boundary."

Littlewood's Inequality. *If φ is a holomorphic self-map of U, then for each $w \in U\backslash\{\varphi(0)\}$,*

$$N_\varphi(w) \leq \log \left| \frac{1 - \overline{w}\varphi(0)}{w - \varphi(0)} \right|.$$

Proof. The right-hand side of the desired inequality is $-\log|\alpha_w(\varphi(0))|$, so it is non-negative. If $w \notin \varphi(U)$ then the left side is zero, in which case there is nothing to prove. For $w \in \varphi(U)$, with $w \neq \varphi(0)$, write $n(r, w)$ for the number of terms of $\varphi^{-1}\{w\} = \{z_n(w)\}$ that lie in the closed disc $r\overline{U}$ ($0 \leq r < 1$). Apply Jensen's Formula (§7.3) to the function $f = \alpha_w \circ \varphi$ (which has $\varphi^{-1}\{w\}$ as its zero-sequence) to get

$$\sum_{n=1}^{n(r,w)} \log \frac{r}{|z_n(w)|} = \frac{1}{2\pi} \int_{-\pi}^{\pi} \log|\alpha_w(\varphi(re^{i\theta}))|\, d\theta + \log \frac{1}{|\alpha_w(\varphi(0))|}$$

for each $0 \leq r < 1$. Since $|\alpha_w \circ \varphi| < 1$ at each point of U, the integral on the right is negative, hence

$$\sum_{n=1}^{n(r,w)} \log \frac{r}{|z_n(w)|} < \log \frac{1}{|\alpha_w(\varphi(0))|}. \tag{16}$$

If $\varphi^{-1}\{w\}$ is finite, let r tend to 1 on the left-hand side of this inequality to obtain

$$N_\varphi(w) = \sum_n \log \frac{1}{|z_n(w)|} \leq \log \frac{1}{|\alpha_w(\varphi(0))|}, \tag{17}$$

which is the desired inequality. If, on the other hand, $\varphi^{-1}\{w\}$ is infinite, then for each positive integer N we can choose an $0 < R < 1$ so that $n(R, w) \geq N$. Then for $R \leq r < 1$, inequality (16) yields

$$\sum_{n=1}^{N} \log \frac{r}{|z_n(w)|} \leq \sum_{n=1}^{n(r,w)} \log \frac{r}{|z_n(w)|} \leq \log \frac{1}{|\alpha_w(\varphi(0))|}.$$

On the left side of this chain of inequalities, first let r tend to 1, and then send N to ∞. Once again the result is (17). $\square$

Corollary. *For any holomorphic self-map* φ *of* U,

(a) $N_\varphi(w) = O\left(\log \frac{1}{|w|}\right)$ *as* $|w| \to 1-$.

(b) *If* $\varphi(0) = 0$ *then, more precisely,* $N_\varphi(w) \le \log \frac{1}{|w|}$ *for each* $w \in U$.

Proof. Part (b) is just Littlewood's Inequality with $\varphi(0) = 0$. For part (a), recall the identity

$$1 - \left|\frac{p - w}{1 - \overline{w}p}\right|^2 = \frac{(1 - |p|^2)(1 - |w|^2)}{|1 - \overline{w}p|^2} \qquad (w, p \in U) \qquad (18)$$

that we employed in our work on the Invariant Schwarz Lemma (equation (2) of §4.3). Equation (18), Littlewood's Inequality, and the fact that $(x - 1)^{-1} \log x \to 1$ as $x \to 1$, quickly yield

$$\limsup_{|w| \to 1-} \frac{N_\varphi(w)}{\log \frac{1}{|w|}} \le \frac{1 + |\varphi(0)|}{1 - |\varphi(0)|},$$

from which follows (a). $\qquad\qquad\qquad\qquad\qquad\qquad\qquad\qquad\qquad\square$

Remarks. (a) As we pointed out earlier, this Corollary, along with the Littlewood-Paley identity and the change-of-variable formula "explains" why all composition operators are bounded on H^2. The more precise statement, part (b), provides a similar explanation for Littlewood's original Subordination Theorem (§1.3), which asserts that C_φ is *contractive* whenever φ fixes the origin.

(b) Note also that part (b) of the Corollary is actually a generalization of the Schwarz Lemma! Indeed, upon exponentiating both sides of the inequality we obtain:

If $\varphi(0) = 0$ *then for each* $w \in \varphi(U)$, *the product of all the terms in* $\varphi^{-1}\{w\}$—*even counting multiplicities—has modulus* $\ge |w|$.

By contrast, the Schwarz Lemma merely asserts that each *individual* term in $\varphi^{-1}\{w\}$ has modulus $\ge |w|$.

10.5 Proof of Sufficiency

The argument here is entirely similar to that of §3.3. We are assuming (7):

$$\lim_{|w| \to 1-} N_\varphi(w) = o\left(\log \frac{1}{|w|}\right) \qquad \text{as} \quad |w| \to 1-,$$

and wish to prove that C_φ is compact on H^2.

For this, fix a sequence of functions $\{f_n\}$ in the unit ball of H^2 that converges to zero uniformly on compact subsets of U. By the Weak Convergence Theorem of §2.4 it is enough to show that $\|C_\varphi f_n\| \to 0$.

Let $\varepsilon > 0$ be given, and use the hypothesis on N_φ to choose $0 < r < 1$ so that

$$N_\varphi(w) < \varepsilon \log \frac{1}{|w|} \quad \text{whenever} \quad r \le |w| < 1. \tag{19}$$

Since $f_n \to 0$ uniformly on compact sets, we can choose n_ε so that $|f_n| < \sqrt{\varepsilon}$ on $rU \cup \{\varphi(0)\}$ whenever $n > n_\varepsilon$. Thus for each such n we obtain from the change-of-variable formula (4),

$$\|C_\varphi f_n\|^2 = |f_n(\varphi(0))|^2 + \int_{rU} + \int_{U \setminus rU} |f_n'(w)|^2 N_\varphi(w)\, dA(w).$$

$$< \varepsilon + \varepsilon \int_{rU} N_\varphi(w)\, dA(w) + \varepsilon \int_{U \setminus rU} |f_n'(w)|^2 \log \frac{1}{|w|}\, dA(w)$$

$$\le \varepsilon + \varepsilon \int_U N_\varphi(w)\, dA(w) + \varepsilon \int_U |f_n'(w)|^2 \log \frac{1}{|w|}\, dA(w)$$

$$= \varepsilon + \frac{\varepsilon}{2} \|z\|^2 + \frac{\varepsilon}{2} \left(\|f_n\|^2 - |f_n(\varphi(0))|^2 \right)$$

$$\le \varepsilon + \frac{\varepsilon}{2} + \frac{\varepsilon}{2} = \varepsilon,$$

where in the next-to-last line we have used formula (4) twice—the first time with $f(z) = z$, and the second time at full strength—and in the last line we finally used the fact that $\|f_n\| \le 1$ for each n. Thus $\|C_\varphi f_n\| \to 0$, which establishes the compactness of C_φ on H^2. $\qquad \square$

10.6 Averaging the Counting Function

The example presented in §10.2 shows that the operator-theoretic method used to prove the "necessity" part of the Univalent Compactness Theorem falls short in the general case. In its place we are going to use a function-theoretic argument based on a special averaging property of the Nevanlinna Counting Function. Ultimately, this property will allow us to estimate the values of the counting function from the action of the corresponding composition operator on some natural test functions. The experienced reader will recognize the hand of subharmonicity at work throughout this section.

The sub-averaging property. *Suppose ψ is a holomorphic self-map of U with $\psi(0) \neq 0$. If $0 < R < |\psi(0)|$, then*

$$N_\psi(0) \leq \frac{1}{R^2} \int_{RU} N_\psi \, dA. \tag{20}$$

Remark. Since $R^2 = A(RU)$ this result asserts that the average of the counting function over the disc RU dominates its value at the center. The proof we are going to give works with minor modifications to yield the same result for any disc in U that does not contain $\psi(0)$.

Proof. In Jensen's Formula ((9) of §7.3), the terms of the sum on the left side of the equation are all positive, so there results this inequality for functions f holomorphic on U with $f(0) \neq 0$:

$$\log |f(0)| \leq \frac{1}{2\pi} \int_{-\pi}^{\pi} \log |f(re^{i\theta})| \, d\theta \qquad (0 \leq r < 1).$$

So if $w \in U$, then upon setting $f(z) = z - w$ in this inequality we obtain

$$\log |w| \leq \frac{1}{2\pi} \int_{-\pi}^{\pi} \log |re^{i\theta} - w| \, d\theta, \tag{21}$$

for each $0 \leq r < 1$. (The subharmonically enlightened will recognize that we are just listing consequences of the subharmonicity of the logarithm of the magnitude of an analytic function.) Note that the heart of (21) is the case $r \geq |w|$, since otherwise the function being integrated is harmonic in a neighborhood of $r\overline{U}$, in which case the mean value property of harmonic functions yields (21), with equality.

Now integrate (21) over the interval $[0, R]$ with respect to the measure $2R^{-2}r \, dr$ to obtain

$$\log |w| \leq \frac{1}{R^2} \int_{RU} \log |z - w| \, dA(w). \tag{22}$$

As is our custom, for $w \in U\backslash\{\psi(0)\}$ we employ the notation $\{z_n(w)\}$ for the points of $\psi^{-1}\{w\}$, listed in order of increasing moduli, and repeated according to multiplicity, and we write $n(r, w)$ for the number of terms of this sequence that have modulus $\leq r$. Let

$$N_{\psi,r}(w) \stackrel{\text{def}}{=} \sum_{n=1}^{n(r,w)} \log \frac{r}{|z_n(w)|}$$

for $0 \leq r < 1$. Then Jensen's Formula, with $f = \psi - w$, reads:

$$N_{\psi,r}(w) = \frac{1}{2\pi} \int_{-\pi}^{\pi} \log |\psi(re^{i\theta}) - w| \, d\theta - \log |\psi(0) - w| \tag{23}$$

for $0 \leq r < 1$. Integrate both sides of this identity with respect to the probability measure $R^{-2} dA(w)$, and use Fubini's Theorem to get

$$\frac{1}{R^2} \int_{RU} N_{\psi,r}(w) \, dA(w)$$

$$= \frac{1}{2\pi} \int_{-\pi}^{\pi} \left\{ \frac{1}{R^2} \int_{RU} \log |\psi(re^{i\theta}) - w| \, dA(w) \right\} d\theta - \log |\psi(0)|,$$

where the fact that $|\psi(0)| > R$ allows the mean value property of harmonic functions to be used to get the second term on the right (cf. the comment following (21) above).

Now use inequality (22) on the inner integral on the right, with $z = \psi(re^{i\theta})$, to obtain for each $0 \leq r < 1$:

$$\frac{1}{R^2} \int_{RU} N_{\psi,r}(w) \, dA(w) \geq \frac{1}{2\pi} \int_{-\pi}^{\pi} \log |\psi(re^{i\theta})| \, d\theta - \log |\psi(0)|$$

$$= N_{\psi,r}(0) \qquad \text{(by Jensen's formula)}.$$

The proof is completed by observing that for each $w \in U$,

$$N_{\psi,r}(w) \nearrow N_\psi(w) \quad \text{as} \quad r \nearrow 1,$$

so the desired inequality on N_ψ follows from the one above for $N_{\psi,r}$ and the Lebesgue Monotone Convergence Theorem.   □

Remark. Equation (23) above shows that $N_{\psi,r}(w)$ is subharmonic on $U \backslash \{\psi(0)\}$, and this gives a more sophisticated way of deriving the sub-averaging inequality (20) for these "partial counting functions." However N_φ itself, although it also has the sub-averaging property, need *not* be subharmonic. Indeed N_φ, being an increasing limit of the continuous functions $N_{\varphi,r}$, is lower semicontinuous, but subharmonicity demands *upper* semicontinuity. This having been noted, we hasten to add that more sophisticated analysis shows that N_φ fails to be subharmonic only on a set of logarithmic capacity zero (see [ESS '85, page 130] for further references and details).

10.7 Proof of Necessity

In this section we assume that C_φ is compact on H^2, and prove that $N_\varphi(w) = o(\log(1/|w|))$ as $|w| \to 1-$, or what is the same:

$$\lim_{|w| \to 1-} \frac{N_\varphi(w)}{1 - |w|} = 0. \tag{24}$$

For this we need a simple transformation formula for the counting function.

Lemma. For $p \in U$ let α_p denote the special automorphism of U that interchanges p and the origin. Then

$$N_\varphi(\alpha_p(w)) = N_{\alpha_p \circ \varphi}(w)$$

for every $w \in U$.

Proof. Since α_p is its own compositional inverse, we see that for each complex number w, the functions $\varphi - \alpha_p(w)$ and $\alpha_p \circ \varphi - w$ share the same zero-sequence. The result follows immediately.   □

For $p \in U$ recall the "normalized reproducing kernel"

$$f_p(z) = \frac{\sqrt{1 - |p|^2}}{1 - \bar{p}z}$$

that occurred in both §2.5, and the necessity part of the Univalent Compactness Theorem (§3.5). Since $\|f_p\| = 1$ for each p, and $f_p \to 0$ uniformly on compact subsets of U as $|p| \to 1-$, it follows from the compactness of C_φ and the Weak Convergence Theorem (§2.4) that

$$\lim_{|p| \to 1-} \|C_\varphi f_p\| = 0.$$

(In Chapter 3 we used the *adjoint* of C_φ instead of the operator itself in this equation.) Our strategy here will be to apply the change-of-variable formula (4) to the functions f_p, and then use the Lemma and the sub-averaging property (20) of the last section to get information about the values of the counting function. From (4) we obtain:

$$\|C_\varphi f_p\|^2 \;\geq\; 2 \int_U |f_p'(w)|^2 N_\varphi(w)\, dA(w)$$

$$= \; 2 \int_U \frac{(1 - |p|^2)|p|^2}{|1 - \bar{p}w|^4} N_\varphi(w)\, dA(w)$$

$$= \; \frac{2|p|^2}{1 - |p|^2} \int_U |\alpha_p'(w)|^2 N_\varphi(w)\, dA(w).$$

Now make the substitution $W = \alpha_p(w)$ in the last integral, and use the Lemma:

$$\|C_\varphi f_p\|^2 \;\geq\; \frac{2|p|^2}{1 - |p|^2} \int_U N_\varphi(\alpha_p(W))\, dA(W)$$

$$= \; \frac{2|p|^2}{1 - |p|^2} \int_U N_{\alpha_p \circ \varphi}(W)\, dA(W)$$

$$\geq \; \frac{2|p|^2}{1 - |p|^2} \int_{\frac{1}{2}U} N_{\alpha_p \circ \varphi}(W)\, dA(W).$$

Apply the sub-averaging inequality (20) to the last integral, with $\psi = \alpha_p \circ \varphi$, and use the Lemma one last time to finish the estimate:

$$\|C_\varphi f_p\|^2 \;\geq\; 4 \cdot \frac{2|p|^2}{1 - |p|^2} N_{\alpha_p \circ \varphi}(0)$$

$$= \;\frac{8|p|^2}{1 + |p|} \cdot \frac{N_\varphi(p)}{1 - |p|}.$$

Note that in the first line of the above display, the application of the sub-averaging property (20) over the disc $\frac{1}{2}U$ requires $|\alpha_p(\varphi(0))| > 1/2$. But $|\alpha_p(\varphi(0))| \to 1$ as $|p| \to 1-$, so this happens for all points p sufficiently close to ∂U. Hence for all such p,

$$\|C_\varphi f_p\|^2 \geq \mathrm{const.}\; \frac{N_\varphi(p)}{1 - |p|}.$$

As we saw above, the compactness of C_φ forces $\|C_\varphi f_p\|$ to zero as $|p| \to 1-$, so the last inequality yields the desired estimate (24) on the counting function.   □

10.8   Exercises

1. Prove that the Frostman condition (13) of §10.2 is also *sufficient* for the Blaschke product B to have an angular derivative at ζ.

   Suggestion: Use the result of Exercise 12 of §4.8.

2. Suppose φ is a holomorphic self-map of U that has an angular derivative at $\zeta \in \partial U$. Show that the zero-sequence $\{b_n\}$ of φ satisfies the Frostman condition (13).

   Suggestion: Write $\varphi = B_n \varphi_n$ where B_n is an appropriate finite product of special automorphisms, and use the Julia-Carathéodory Theorem along with the result of Exercise 9 of §4.8.

3. Show that

   $$N_\varphi(w) = \int_0^1 \frac{n(r, w)}{r}\, dr$$

   where φ is a holomorphic self-map of U and $n(r, w)$ is the number of terms of the inverse-image sequence $\varphi^{-1}\{w\}$ in the closed disc $r\overline{U}$.

4. Show that if $f \in H^2$ is not identically zero, then

   $$n(r) \log \frac{1}{r} \to 0 \quad \text{as} \quad r \to 1-.$$

5. Use the previous problem to show that if B is a Blaschke product then
$$\int_0^{2\pi} \log|B(re^{i\theta})|\,d\theta \to 0 \quad \text{as} \quad r \to 1-.$$
(See the *Notes* below for the significance of this result.)

6. Show that if $f \in H^2$ then $f = BF$ where B is a Blaschke product, $F \in H^2$, with $\|F\| = \|f\|$, and F vanishes nowhere on U.

7. Using the problem above, show that $f \in H^2$ if and only if $f = gh$ where g and h belong to H^1 (see §6.4 for the definition of H^1).

8. Show that if $f \in H^2$ then $f = g + h$ where both g and h belong to H^2, neither has any zeros in U, and both have norm $\leq \|f\|$.

9. Show that C_φ satisfies the Hilbert-Schmidt criterion of §2.3 if and only if
$$\int_U \frac{N_\varphi(w)}{(1-|w|)^3}\,dA(w) < \infty$$
(cf. Exercise 2 of §3.7). *Suggestion:* Recall that the Hilbert-Schmidt criterion is equivalent to the convergence of the sum $\sum \|\varphi^n\|^2$. Use the change-of-variable formula (4) to compute the norm of each term as an area integral involving the counting function.

The problems below solve the compactness problem for arbitrary composition operators on the Bergman space A^2—the space of holomorphic functions f that belong to $L^2(U, dA)$ (see Exercise 4 of §1.4). According to Exercise 10 of §3.7, just as for H^2, the angular derivative criterion is necessary for compactness of C_φ on A^2, even when φ is not univalent. The problems below show that, by contrast with the H^2 case, the angular derivative criterion is, for general composition operators, also *sufficient* for compactness on A^2 (for references, see §3.8).

10. Show that the norm in A^2 can be estimated as follows:
$$\|f\|^2 \approx |f(0)|^2 + \int_U |f'(z)|^2 \left(\log \frac{1}{|z|}\right)^2 dA(z)$$
(cf. Exercise 9 of §3.7).

11. Show that if $f \in A^2$ and φ is any holomorphic self-map of U, then the Bergman space norm of $f \circ \varphi$ is estimated by:
$$\|f \circ \varphi\|^2 \approx |f(\varphi(0))|^2 + \int_U |f'(w)|^2 M_\varphi(w)\,dA(w),$$
where M_φ is defined like N_φ, except that you sum $(\log(1/|z|))^2$ instead of $\log(1/|z|)$.

12. Prove Littlewood's inequality for M_φ when $\varphi(0) = 0$:

$$M_\varphi(w) \le \left(\log \frac{1}{|w|}\right)^2 \qquad (w \in U).$$

Suggestion: Use the original Littlewood inequality, along with an inequality on series.

What happens to this inequality when $\varphi(0) \ne 0$?

13. Use the result of the last problem to show that every composition operator is bounded on A^2. (Note that it is enough to assume $\varphi(0) = 0$ for this.)

14. Show that a *sufficient* condition for C_φ to be compact on A^2 is that

$$M_\varphi(w) = o\left(\left(\log \frac{1}{|w|}\right)^2\right) \qquad \text{as } |w| \to 1-. \qquad (25)$$

15. Show that if φ has an angular derivative at no point of ∂U, then condition (25) above is satisfied, hence C_φ is compact on A^2.

Suggestion: If $\varphi(0) = 0$, then the Schwarz Lemma implies that $|w| \le |z|$ for each $z \in \varphi^{-1}\{w\}$. Thus, given $\varepsilon > 0$ you can use the angular derivative criterion to insure that if w is close enough to ∂U, then $\log(1/|z|) < \varepsilon \log(1/|w|)$ for each $z \in \varphi^{-1}\{w\}$. Then use the original Littlewood inequality to finish the job.

10.9 Notes

The Compactness Theorem. This result and its proof are taken from [Sh1 '87], where a more precise result is obtained:

The distance, in the operator norm, from a composition operator C_φ to the subspace of compact operators is

$$\limsup_{|w| \to 1-} \sqrt{\frac{N_\varphi(w)}{\log \frac{1}{|w|}}} .$$

This result is applied to the study of spectra in [CwM '92]. An elementary result of this sort has already occurred here in §2.6, Exercise 3, where a formula is given for the distance from a *diagonal* operator to the space of compacts. Curiously, there is no expression known for the *norm* of a general composition operator; in fact the problem seems quite intractable (for an idea of how complicated this can get, see [Cwn '88] for the computation of the norm of C_φ, where $\varphi(z) = az + b$, with $|a| + |b| \le 1$).

The Counting Function. For Bergman spaces, the application of counting function methods to the problem of determining compactness for composition operators (Exercises 10 through 15 above) occurs in [Sh1 '87], where the same result is obtained for various weighted Bergman spaces. More generally, the work of §10.3 suggests that any Hilbert space of analytic functions defined by integrating $|f'|^2$ against a suitable measure on U will have its own counting function, and this will control questions of boundedness and compactness for composition operators on that space. This theme is developed further in [KrM '90] and [KrM '92].

The importance of the space H^2 as a setting for the theory of composition operators is further underscored by the fact that the counting function associated with H^2 is the classical one introduced by Nevanlinna. The Nevanlinna Counting Function is a fundamental object in the study of the distribution of values of holomorphic functions (see, for example, [Nvl '53], or [Hmn '64]).

The application of the Nevanlinna Counting Function to composition operators has its roots in a remarkable formula of C.S. Stanton for the integral means of a subharmonic function in the disc. Stanton showed that if u is a positive subharmonic function on U, and φ is a holomorphic self-map of U, then for $0 < r < 1$,

$$\frac{1}{2\pi} \int_{-\pi}^{\pi} u(re^{i\theta})\, d\theta = u(\varphi(0)) + \int N_{\varphi,r}(w)\, d\mu(w),$$

where μ is the *Riesz mass* of u. The change-of-variable formula (4) on which the work of this chapter was based comes from the special case $u = |f'|^2$ of Stanton's formula; For another application to composition operators, see [SS2 '90].

The Hilbert-Schmidt criterion of Exercise 9 above can also be derived from the one presented in §2.3 via Stanton's formula.

Stanton's formula first appeared in his thesis [Stn '82], where it was applied to give unified proofs of a number of important classical function-theoretic inequalities. These results were first published in [ESS '85]. For further applications, see [Stn '86].

Schatten classes. The *Schatten p-class* S_p on a Hilbert space is the collection of compact operators T such that the eigenvalues of T^*T form a sequence in $\ell^{p/2}$. It turns out that S_2 is just the Hilbert-Schmidt class. Questions about the membership of composition operators in Schatten classes were first raised in [STa '73], where it was shown that if $\varphi(U)$ lies in an inscribed polygon, then $C_\varphi \in S_p$ for all $p > 0$ (the $p = 2$ case was proved here in §2.3).

The characterization of Hilbert-Schmidt composition operators in terms of the Nevanlinna Counting Function, given in Exercise 9 above, can be rewritten:

$$\int_U \frac{N_\varphi(w)}{\log \frac{1}{|w|}}\, d\nu(w) < \infty,$$

where $d\nu(w)$ is the hyperbolic area measure introduced in Exercise 10 of §9.10. A heuristic interpolation between this result and the Compactness Theorem, leads one to conjecture that for $2 \le p < \infty$

C_φ belongs to the Schatten p-class if and only if

$$\left(\frac{N_\varphi(w)}{\log \frac{1}{|w|}} \right) \in L^{p/2}(\nu).$$

Luecking and Zhu [LZh '92] recently proved that this is true, and even holds for the full range $0 < p < \infty$.

Frostman's Theorem. This result, proved in §10.2 and Exercise 1 above, is from [Fmn '42], where it is also shown that if, in the condition (13), one replaces the exponent "2" in the denominator of each term of the sum by "1," then the resulting weaker condition is necessary and sufficient for the Blaschke product and all its partial products to have radial limits of modulus 1 at ζ. The fact that this condition must hold for the zeros of any holomorphic self-map of U that has an angular derivative (Exercise 2 above) is from [Cth '54, §301–§304].

Equality in Littlewood's Inequality. Since Littlewood's Inequality is a generalization of the Schwarz Lemma (§10.4), it is natural to ask about the case of equality here. It turns out that the situation is more complicated than might be expected. It follows directly from the derivation of Littlewood's Inequality that there is equality at a point $w \in U$ if and only if

$$\lim_{r \to 1-} \int_{-\pi}^{\pi} \log |\alpha_w(\varphi(re^{i\theta}))| \, d\theta = 0,$$

and it is well-known that this happens if and only if the function $\alpha_w \circ \varphi$ is a Blaschke product (see [CdL '66, Th. 2.12, page 32]; the easy part of this has occurred here as Exercise 5 above). Now another result of Frostman asserts that if φ is any *inner function* (i.e., $|\varphi| < 1$ at each point of U, and $|\varphi^*| = 1$ a.e. on ∂U), then $\alpha_w \circ \varphi$ is a Blaschke product for every $w \in U$ that lies outside a (possibly empty) set of logarithmic capacity zero ([CdL '66, Theorem 2.15, page 37]). Putting it all together we obtain the equivalence of the following assertions (cf. [Lto '53], [Ltw '25], [Sh1 '87]):

- There is equality in Littlewood's Inequality for some $w \in U$.

- There is equality in Littlewood's Inequality for all $w \in U$ outside a set of logarithmic capacity zero.

- φ is an inner function.

Note that these observations, along with the fundamental integral formulas (3) and (4) of §10.1 show that if φ is an inner function that vanishes at the origin, then C_φ takes H^2 *isometrically* into itself, a result first proved, by different methods, by Nordgren [Ngn '68].

Epilogue

I originally planned to conclude this book with a brief discussion of unsolved problems and promising research directions. However, a couple of futile attempts made it clear that new directions should be suggested, not by my own pronouncements, but by each reader's individual interests.

Clearly, any question about operators yields a question about composition operators. The point is that the good questions relate naturally to both the function theory of the inducing map, and the investigator's particular strengths in analysis. By developing a few threads of interest to me —the studies of compactness and cyclicity, along with the classical function theory that naturally attaches itself these problems—I hope to have demonstrated how one such program can evolve.

Unfortunately the small size and narrow focus of this book require that more be omitted than included. While the *Notes* at the end of each chapter point out some further directions, many gaps remain in the record. Thus, I would like to close this exposition by indicating a few more avenues of inquiry that have proven interesting, but did not find their way into the body of the text.

Operator algebras and invariant subspaces. The algebras generated by various classes of composition operators have been studied by Cima and Wogen [CWo '74] as well as Nordgren, Rosenthal and Wintrobe [NRW '87]. Guyker [Gkr '89] classified the reducing subspaces of certain composition operators. For other results on invariant subspaces, see [Ngn '68] and §6 of [NRW '87].

Closed Range. Cima, Thomson, and Wogen characterized the closed-range composition operators on H^2, phrasing their result in terms of the boundary behavior of the inducing function [CTW '74]. They also characterized the Fredholm composition operators as precisely the invertible ones (the ones induced by disc automorphisms). Recently, Zorboska [Zrb '92] has characterized the closed-range composition operators in terms of the behavior of the Nevanlinna Counting Function.

Small spaces. In Chapter 1 we mentioned Alp'ar's result that disc automorphisms that do not fix the origin cannot take $\ell^1(U)$ into itself (§1.4, Exercise 10, and §1.5). This suggests the study of composition operators on spaces of analytic functions smaller than H^2. For a sampling of results in this setting see [MlS '86], [Mcl '87], [Rn '78], [Sh2 '87], [Zrb '89], [Mdn '92].

Semigroups of composition operators. Investigation of this area began in 1972 with J.A. Deddens's observation that certain composition operators of the form C_{az+b} can be represented as functions of the adjoint of the classical Cesàro operator [Dns '72]. In [Cwn '84] Cowen studied the connection between the the Cesàro operator and certain semigroups of composition operators. Berkson and Porta initiated a different line of investigation into composition operator semigroups [BPo '78], emphasizing function-theoretic properties of the infinitesimal generators of these semigroups, and their work was carried forward by Siskakis [Sk1 '87], [Sk2 '87] and Aleman [Amn '90].

The space of composition operators. Motivated by his work with Porta on semigroups, Berkson proved that each composition operator induced by an inner function is isolated, in the operator-norm topology, from every other composition operator [Bsn '81]. Shapiro and Sundberg related isolation phenomena with extreme points of H^∞ (only extreme points can induce such isolated composition operators, and all "sufficiently regular" univalent extreme points actually do induce isolated operators) [SS1 '90], while MacCluer studied the connection between angular derivatives and components in the space of composition operators [Mlr '89]. Jarchow, Hunziger, and Maschioni have studied similar problems for other situations, e.g., the class of Hilbert-Schmidt composition operators, in the topology induced by the Hilbert-Schmidt norm [JHM '90].

Composition operators on L^1 and M. Here L^1 denotes the Lebesgue space of the unit circle, defined with respect to normalized arc-length measure, and M is the Banach space of complex Borel measures on the circle. It is not difficult to see that for φ a holomorphic self-map of U, if u is a harmonic function on U that is the Poisson integral a measure on ∂U, then the same is true of $u \circ \varphi$. Thus the composition operator C_φ can be defined on M, and it is bounded there. Sarason proved that C_φ acts boundedly on L^1 as well [Ssn '90]; he showed that for composition operators, compactness on L^1 is equivalent to compactness on M, and that this is equivalent to weak compactness on either space. Sarason then asked if the compact composition operators on L^1 must coincide with those on H^2, a question answered affirmatively in [SS2 '90]. (One could also pose Sarason's question for $1 < p < \infty$, but here the answer is trivially "yes," since if $\varphi(0) = 0$, then

L^p has an obvious direct sum decomposition into (boundary) $H^p \bigoplus \overline{zH^p}$, both subspaces being invariant under C_φ.) Questions of compactness for composition operators on M and L^1 arise in connection with problems in prediction theory (see [Ssn '90]).

References

[Abt '91]   M. Abate, *Iteration theory, compactly divergent sequences and commuting holomorphic maps*, Ann. della Scuola Normale Superiore di Pisa (4) **18** (1991), 167–191.

[Ahl '66]   L.V. Ahlfors, *Complex Analysis*, McGraw-Hill, New York, 1966.

[Akd '87]   J.Akeroyd, *Polynomial approximation in the mean with respect to harmonic measure*, Trans. Amer. Math. Soc. **303** (1987), 193–199.

[Akd '92]   J.Akeroyd, *Polynomial approximation in the mean with respect to harmonic measure II*, Michigan Math. J. **39** (1992), 35–40.

[AKS '91]   J. Akeroyd, D. Khavinson, and H.S. Shapiro, *Remarks concerning cyclic vectors in Hardy and Bergman spaces*, Michigan Math. J. **38** (1992), 191–205.

[Amn '90]   A. Aleman, *Compactness of resolvent operators generated by a class of composition semigroups on H^p*, J. Math. Anal. Appl. **147** (1990), 171–179.

[Alp '60]   L. Al'par, *Egyes hatbánsyorok abazohit konvergenciája a konvergencia kôr kerûleten*, Mat. Lapok **11** (1960), 312–322.

[BkP '79]   I.N. Baker and Ch. Pommerenke, *On the iteration of analytic functions in a half-plane II*, J. London Math. Soc. (2) **20** (1979), 255–258.

[Br1 '92]   H. S. Bear, *Distance decreasing functions on the hyperbolic plane*, Michigan Math. J. **39** (1992), 271–279.

[Br2 '92]   H.S. Bear, *A new proof of the Julia-Carathéodory Theorems*, preprint, 1992.

[Brd '82]   A.F. Beardon, *The Geometry of Discrete Groups*, Springer-Verlag, New York, 1982.

[Brd '90]   A.F. Beardon, *Iteration of contractions and analytic maps*, J. London Math. Soc. (2) **41** (1990), 141–150.

[Brd '91]   A.F. Beardon, *Iteration of Rational Functions*, Springer-Verlag, New York, 1991.

[Bmy '86]   B. Beauzamy, *Un opérateur sur l'espace de Hilbert, dont tous les polynômes sont hypercycliques*, C.R. Acad. Sci. Paris , Ser. I**303** (1986), 923–927.

[BPo '78]   E. Berkson and H. Porta, *Semigroups of analytic functions and composition operators*, Michigan Math. J. **25** (1978), 101–115.

[Bsn '81]   E. Berkson, *Composition operators isolated in the uniform operator topology*, Proc. Amer. Math. Soc. **81** (1981), 230–232.

[Brk '29]   G.D. Birkhoff, *Démonstration d'un Théoreme elementaire sur les fonctions entieres*, C.R. Acad. Sci. Paris **189** (1929), 473–475.

[Bdn '87]   P.S. Bourdon, *Density of the polynomials in Bergman spaces*, Pacific J. Math. **130** (1987), 215–221.

[Bdn '92]   P.S. Bourdon, *Invariant manifolds of hypercyclic vectors*, Proc. Amer. Math. Soc. (to appear).

[BoS '90]   P.S. Bourdon and J. H. Shapiro, *Cyclic composition operators on H^2*, Proc. Symp. Pure Math., **51**, Part 2 (1990), 43–53.

[BoS '93]   P.S. Bourdon and J.H. Shapiro, *Cyclic properties of composition operators*, preprint 1993.

[Brk '81]   R. B. Burckel, *Iterating analytic self-maps of discs*, Amer. Math. Monthly **88** (1981), 396–407.

[Bdz '86]   K. Burdzy, *Brownian excursions and minimal thinness, Part III*, Math. Z. **192** (1986), 89–107.

[C-W '91]   R.K. Campbell-Wright, *Equivalent composition operators*, Integral Eqns. and Op. Th. **14** (1991) 71–97.

[C-W '93]   R.K. Campbell-Wright, *Similar compact composition operators*, Acta. Sci. Math. (Szeged), to appear.

[Cth '29]   C. Carathéodory, *Über die Winkelderivierten von beschränkten Analytischen Funktionen*, Sitzungsber. Preuss. Akad. Viss. Berlin, Phys.- Math. Kl. (1929), 39–54.

[Cth '54]   C. Carathéodory, *Theory of Functions*, Vol. 2, Chelsea, New York, 1954.

[Crl '88]   T.F. Carroll, *A classical proof of Burdzy's theorem on the angular derivative*, J. London Math. Soc. **38** (1988), 423–441.

[Cgn '71]   J. Caughran, *Polynomial approximation and spectral properties of composition operators on H^2*, Indiana Univ. Math. J. **21** (1971), 81–84.

[CSc '75]   J. Caughran and H.J. Schwartz, *Spectra of compact composition operators*, Proc. Amer. Math. Soc. **51** (1975), 127–130.

[ChS '91]   K.C. Chan and J.H. Shapiro, *The cyclic behavior of translation operators on Hilbert spaces of entire functions*, Indiana Univ. Math. J. **40** (1991), 1421–1449.

[ChB '90]   R.V. Churchill and J.W. Brown, *Complex Variables and Applications*, fifth ed. McGraw-Hill, New York, 1990.

[CTW '74]   J.A. Cima, J. Thomson, and W. Wogen, *On some properties of composition operators*, Indiana Univ. Math. J. **24** (1974), 215–220.

[CWo '74]   J.A. Cima and W.Wogen, *On algebras generated by composition operators*, Canadian J. Math. **26** (1974), 1234–1241.

[CdL '66]   E.F. Collingwood and A.J. Lohwater, *The Theory of Cluster Sets*, Cambridge University Press, 1966.

[Cwn '81]   C.C. Cowen, *Iteration and solution of functional equations for functions analytic in the unit disc*, Trans. Amer. Math. Soc. **256** (1981), 69–95.

[Cwn '83]   C.C. Cowen, *Composition operators on H^2*, J. Operator Th. **9** (1983), 77-106.

[Cwn '84]   C.C. Cowen, *Subnormality of the Cesàro operator and a semigroup of composition operators*, Indiana Univ. Math. J. **33** (1984), 305–318.

[Cwn '88]   C.C. Cowen, *Linear fractional composition operators on H^2*, Integral Eqns. Op. Th. **11** (1988), 151–160.

[Cwn '90]   C.C. Cowen, *Composition operators on Hilbert spaces of analytic functions: A status report*, in Proc. of Symp. in Pure Math., **51** Part I (1990), American Math. Soc., Providence, R.I.

[CwK '88]   C.C. Cowen and T.L. Kriete III, *Subnormality and composition operators on H^2*, J. Functional Analysis **81** (1988), 298–319.

[CwM '92]   C.C. Cowen and B.D. MacCluer, *Spectra of some composition operators*, preprint, 1992.

[Dns '72]   J. Deddens, *Analytic Toeplitz and composition operators*, Canadian J. Math. **5** (1972), 859–865.

[Djy '26]   A. Denjoy, *Sur l'iteration des fonctions analytiques*, C.R. Acad. Sci. Paris **182** (1926), 255–257.

[Dvy '89]   R.L. Devaney, *An Introduction to Chaotic Dynamical Systems*, second ed., Addison-Wesley, Reading, Mass., 1989.

[DdS '63]   N. Dunford and J.T. Schwartz, *Linear Operators, Part II*, Wiley, Interscience, New York, 1963.

[Drn '70]   P.L. Duren, *Theory of H^p Spaces*, Academic Press, New York, 1970.

[Enf '87]   P. Enflo, *On the invariant subspace problem for Banach spaces*, Acta Math. **158** (1987), 213–313.

[ESS '85]   M. Essén, D.F. Shea, and C.S. Stanton, *A value-distribution criterion for the class $L \log L$, and some related questions*, Ann. Inst. Fourier (Grenoble) **35** (1985), 127–150.

[Fan '78]   K. Fan, *Julia's lemma for operators*, Math. Ann. **239** (1978), 241–245.

[Fan '83]   K. Fan, *Iteration of Analytic functions of operators I*, Math. Z. **179** (1982) 293–298; II, Lin. Multilin. Alg. **12** (1982/83) 295–304.

[Frd '29]   L.R. Ford, *Automorphic Functions*, Chelsea, New York, 1929.

[Frd '82]   N. Friedman, *Foundations of Modern Analysis*, Dover, New York, 1982.

[Fmn '42]   O. Frostman, *Sur les produits de Blaschke*, Kungl. Fysiogr. Sälsk. i Lund Förh. **12** (1942), 169–182.

[Gai '85]   D. Gaier, *Lectures on Complex Approximaion*, Birkhaüser, Basel 1985.

[GrS '87]   R.M. Gethner and J.H. Shapiro, *Universal vectors for operators on spaces of holomorphic functions*, Proc. Amer. Math. Soc. **100** (1987), 281–288.

[GgP '76]   F.W. Gehring and B.P. Palka, *Quasiconformally homogeneous domains*, J. Analyse Math. **30** (1976), 172–199.

[GyS '91]   G. Godefroy and J.H. Shapiro, *Operators with dense, invariant, cyclic vector manifolds*, J. Functional Analysis **98** (1991), 229–269.

[GlR '84]   K. Goebel and S. Reich, *Uniform Convexity, Hyperbolic Geometry, and Nonexpansive Mappings*, Marcel Dekker, New York, 1984.

[Gbg '62]   J. L. Goldberg, *Functions with positive real part in a half-plane*, Duke Math. J. **29** (1962), 335–339.

[G-E '87]   K-G. Grosse-Erdmann, *Holomorphe Monster und universelle Funktionen*, Mitt. Math. Sem. Giessen **176** (1987), 1–84.

[Gkr '89]   J. Guyker, *On reducing subspaces of composition operators*, Acta Sci. Math. **53** (1989). 369–376.

[Hlz '67]   G. Halasz, *On Taylor series absolutely convergent on the circle of convegence I*, Publ. Inst. Math. Acad. Serbe Sci. **14** (1967), 63–68.

[Hmn '64]   W.K. Hayman *Meromorphic Functions*, Oxford Mathematical Monographs, Clarendon Press, Oxford, 1964.

[HeW '90]   D. A. Herrero and Z. Wang, *Compact perturbations of hypercyclic and supercylic operators*, Indiana Univ. Math. J. **39** (1990), 819–830.

[Hrv '63]   M. Hervé, *Quelques proprietés des applications analytiques d'une boule à m dimensions dans elle-meme*, J. de Math. Pures et appliqueés (9) **42** (1963), 117–147.

[Hzg '88]   G. Herzog, *Universelle Funktionen*, Diplomarbeit, Universität Karlsruhe, 1988.

[Hzr '89]   H. Hunziger, *Kompositionoperatoren auf Klassischen Hardy-räumen*, Thesis, Universität Zurich, 1989.

[HzJ '91]   H. Hunziger and H. Jarchow, *Composition operators which improve integrability*, Math. Nachr. **152** (1991), 83–91.

[JHM '90]   H. Jarchow, H. Hunziger, and V. Mascioni, *Some topologies on the space of analytic self-maps of the unit disc*, in *Geometry of Banach Spaces* (Strobl 1989), London Math. Soc. Lecture Notes #158, 1990, pp 133–148.

[Jrw '92]   H. Jarchow, *Some factorization properties of composition operators*, in *Progress in Functional Analysis*, Bierstedt et. al., editors, pages 405–413, Elsevier 1992.

[Jrw '93]   H. Jarchow, *Absolutely summing composition operators*, Proc. Conf. Functional Analysis, Essen 1991, to appear.

[Jla '20]   G. Julia, *Extension nouvelle d'un lemme de Schwarz*, Acta Math. **42** (1920), 349–355.

[Kwz '75]   H. Kamowitz, *Spectra of composition operators on H^p*, J. Functional Analysis **18** (1975), 132–150.

[Kit '82]   C. Kitai, *Invariant Closed Sets for Linear Operators*, Thesis, Univ. of Toronto, 1982.

[Kgs '84]   G. Königs, *Recherches sur les intégrales de certaines équationes functionelles*, Annales Ecole Normale Superior (3) **1** 1884, Supplément, 3–41.

[Krz '90]   S.G. Krantz, *Complex Analysis: The Geometric Viewpoint*, Carus Math. Monographs #23, Mathematical Ass'n of America, 1990.

[Krz '92]   S.G. Krantz, *Geometric Analysis and Function Spaces*, CBMS Monograph, American Mathematical Society, in preparation.

[KrM '90]   T.L. Kreite III and B.D. MacCluer, *Mean-square approximation by polynomials on the unit disc*, Trans. Amer. Math. Soc. **322** (1990), 1–34.

[KrM '92]   T.L. Kreite III and B.D. MacCluer, *Composition operators on large weighted Bergman spaces*, Indiana Univ. Math. J. **41** (1992) 755–788.

[Kba '83]   Y. Kubota, *Iteration of holomorphic maps of the unit ball into itself*, Proc. Amer. Math. Soc. **88** (1983), 476–485.

[LVn '29]   E. Landau and G. Valiron, *A deduction from Schwarz's Lemma*, J. London Math. Soc. **4** (1929), 162–163.

[Lnr '66]   J. Lehner, *A Short Course in Automorphic Functions*, Holt, Rinehart and Winston, New York, 1966.

[Lto '53]   O. Lehto, *A majorant principle in the theory of functions*, Math. Scand. **1** (1953), 5–17.

[Lnd '15]   E. Lindelöf, *Sur un principe générale de l'analyse et ses applications à la théorie de la représentation conforme*, Acta Soc. Sci. Fennicae **46** (1915), 1–35.

[Ltw '25]   J.E. Littlewood, *On inequalities in the theory of functions*, Proc. London Math. Soc. **23** (1925), 481–519.

[LZh '92]   D. Luecking and K.Zhu, *Composition operators belonging to the Schatten ideals*, American J. Math. **114** (1992), 1127–1145.

[Luh '86]   W. Luh, *Universal approximation properties of overconvergent power series on open sets*, Analysis **6** (1986), 191–207.

[Mlr '83]   B.D. MacCluer, *Iterates of holomorphic self-maps of the unit ball in C^n*, Michigan Math. J. **30** (1983), 79–106.

[Mlr '85]   B.D. MacCluer, *Compact composition operators on $H^p(B_N)$*, Michigan Math. J. **32** (1985), 237–248.

[Mcl '87]   B.D. MacCluer, *Composition operators on S^p*, Houston J. Math. **13** (1987), 245–254.

[Mlr '89]   B.D. MacCluer, *Components in the space of composition operators*, Integral Eqns. Op. Th. **12** (1989), 725–738.

[MlS '86]   B.D. MacCluer and J.H. Shapiro, *Angular derivatives and compact composition operators on the Hardy and Bergman spaces*, Canadian J. Math. **38** (1986), 878–906.

[McL '52]   G.R. MacLane, *Sequences of derivatives and normal families*, J. Analyse Math. **2** (1952), 72–87.

[Mdn '92]   K. Madigan, *Composition operators into Lipschitz type spaces*, Thesis, SUNY at Albany, 1992.

[Mcz '35]   J. Marcinkiewicz, *Sur les nombres dérivés*, Fund. Math. **24** (1935), 305–308.

[Mkv '67]   A.I. Markushevich, *Theory of Functions of a Complex Variable Vol III*, Prentice Hall, 1967 (translated from Russian by R.A. Silverman).

[Mnf '47]   D. Menchoff, *On the partial sums of trigonometric series*,
             (Russian) Math. Sbornik **20** (1947), 197–238.

[Nvl '22]   F. Nevanlinna and R. Nevanlinna, *Über Eigenschaften ana-
             lytischer Funktionen in der Umgebung einer analytischen
             Stelle oder Linie*, Acta Soc. Fenn. **50** (1922), 1–46.

[Nvl '53]   R. Nevanlinna, *Analytic Functions*, Springer, New York,
             1970 (Translation of *Eindeutige Analtyische Funktionen*, sec-
             ond ed., 1953).

[Nwm '61]   M.H. Newman, *Elements of the Topology of Plane Sets of
             Points*, Cambridge Univ. Press, 1961.

[Ngn '68]   E. A. Nordgren, *Composition operators*, Canadian J. Math.
             20 (1968), 442–449.

[NRW '87]   E. Nordgren, P. Rosenthal, and F.S. Wintrobe, *Invertible
             composition operators on H^p*, J. Functional Analysis **73**
             (1987), 324–344.

[Pvn '92]   M. Pavone, *Chaotic composition operators on trees*, Houston
             J. Math., **18** (1992), 47–56.

[Pmk '79]   Ch. Pommerenke, *On the iteration of analytic functions in a
             half-plane I*, J. London Math. Soc. (2) **19** (1979), 439–447.

[Rd '88]   C. Read, *On the invariant subspace problem for Banach
             spaces*, Israel J. Math. **63** (1988), 1–40.

[Rsz '18]   F. Riesz, *Über lineare Funktionalgleichungen*, Acta Math.
             41 1918, 71–98.

[RNg '55]   F. Riesz and B. Sz.-Nagy, Functional Analysis, Ungar, New
             York, 1955 (reprinted by Dover, 1990).

[Rn '78]   R. Roan, *Composition operators on the space of functions
             with H^p derivative*, Houston J. Math. **4** (1978), 423–438.

[RWr '86]   B. Rodin and S.E. Warchawski, *Remarks on a paper of Bur-
             dzy*, J. D'Analyse Math. **46** (1986), 251–260.

[Rlz '69]   S. Rolewicz, *On orbits of elements*, Studia Math. **33** (1969),
             17–22.

[Rdn '87]   W. Rudin, *Real and Complex Analysis*, third ed., McGraw-
             Hill, New York, 1987.

[Rdn '76]   W. Rudin, *Principles of Mathematical Analysis*, third edi-
             tion, McGraw-Hill, New York, 1976.

[Rdn '80] W. Rudin, *Function Theory in the Unit Ball of* C^n, Springer-Verlag, New York, 1980.

[Rdn '91] W. Rudin, *Functional Analysis*, second ed., McGraw-Hill, New York, 1991.

[Rff '66] J.V. Ryff, *Subordinate* H^p *functions*, Duke Math. J. **33** (1966), 347–354.

[Sls '91] H. Salas, *A hypercyclic vector whose adjoint is also hypercyclic*, Proc. Amer. Math. Soc. **112** 1991, 765–770.

[Ssn '88] D. Sarason, *Angular derivatives via Hilbert space*, Complex Variables **10** (1988). 1–10.

[Ssn '90] D. Sarason. *Composition operators as integral operators*, in *Analysis and Partial Differential Equations*, C. Sadowski, ed., Marcel Dekker, Basel, 1990.

[Sdr '30] J. Schauder, *Über lineare, volstetige Funktionaloperationen*, Studia Math. **2** 1951, 183–196.

[Shr '71] E. Schröder, *Über itierte Funktionen*, Math. Ann.**3** (1871), 296–322.

[Schw '69] H.J. Schwartz, *Composition Operators on* H^p, Thesis, Univ. of Toledo, 1968.

[SWa '41] W. Seidel and J. Walsh, *On approximation by Euclidean and non-Euclidean translates of an analytic functions*, Bull. Amer. Math. Soc. **47** (1941), 916–920.

[Srn '56] J. Serrin, *A note on harmonic functions defined in a half-plane*, Duke Math. J. **23** (1956), 523–526.

[STa '73] J.H. Shapiro and P.D. Taylor, *Compact, nuclear, and Hilbert-Schmidt composition operators on* H^2, Indiana Univ. Math. J. **125** (1973), 471–496.

[SSS '92] J.H. Shapiro, W. Smith, and D.A. Stegenga, *Geometric models and compactness of composition operators*, preprint.

[Sh1 '87] J.H. Shapiro, *The essential norm of a composition operator*, Annals of Math. **125** (1987), 375–404.

[Sh2 '87] J.H. Shapiro, *Composition operators on spaces of boundary-regular holomorphic functions*, Proc. Amer. Math. Soc., **100** (1987), 49–57.

[SS1 '90] J.H. Shapiro and C. Sundberg, *Isolation amongst the composition operators*, Pacific J. Math. **145** (1990), 117–152.

[SS2 '90]   J.H. Shapiro and C. Sundberg, *Compact composition oper-ators on L^1*, Proc. Amer. Math. Soc. **108** (1990), 443–449.

[Shd '74]   A.L. Shields, *Weighted shift operators and analytic function theory*, Math. Surveys **13**: *Topics in Operator Theory*, C. Pearcy, ed., American Math. Society 1974, 49–128.

[Sk1 '87]   A.G. Siskakis, *Composition operators and the Cesàro oper-ator on H^p*, J. London Math.soc. (2) **36** (1987) 153–164.

[Sk2 '87]   A.G. Siskakis, *On a class of composition semigroups in Hardy spaces*, J. Math. Anal. Appl. **127** (1987), 122–129.

[Stn '82]   C.S. Stanton, *Riesz mass and growth problems for subhar-monic functions*, Thesis, Univ. of Wisconsin, Madison 1982.

[Stn '86]   C.S. Stanton, *Counting functions and majorization for Jensen measures*, Pacific J. Math. **125** (1986), 459–468.

[Tsj '59]   M. Tsuji, *Potential Theory in Modern Function Theory*, Maruzen, Tokyo, 1959.

[Ulr '83]   D.C. Ullrich, unpublished notes.

[Ulv '72]   P.L. Ul'yanov, *Representation of functions by series and classes $\varphi(L)$*, Russ. Math. Surveys **27** no. 2 (1972), 1–54.

[Vln '31]   G. Valiron, *Sur l'iteration des fonctions holomorphes dans un demi-plan*, Bull des Sci. Math. (2) **55** (1931), 105–128.

[Vnn '75]   S.M. Voronin, *A theorem on the "universality" of the Rie-mann zeta-function*, Izv. Akad. Nauk SSSR Ser. Mat. **39** (1975) 475–486. English translation in: Math. USSR–Izv. **9** (1975), 443–453.

[Yng '86]   R.M. Young, *On Jensen's formula and $\int_0^{2\pi} \log|1 - e^{i\theta}| \, d\theta$*, American Math. Monthly **93** (1986), 44–45.

[Wf1 '26]   J. Wolff, *Sur l'iteration des fonctions borneés*, C.R. Acad. Sci. Paris **182** (1926), 200–201.

[Wf2 '26]   J. Wolff, *Sur une généralisation d'un théorème de Schwarz*, C.R. Acad. Sci. Paris **182** (1926), 918–920, and **183** (1926), 500–502.

[Wgn '90]   W. Wogen, *Composition operators acting on spaces of holo-morphic functions on domains in $\mathbf{C}^n$*, Proc. Symp. Pure Math., **51**, Part 2 (1990), 361–366.

[Zrb '87]   N. Zorboska, *Composition operators on weighted Hardy spaces*, Thesis, University of Toronto, 1987.

[Zrb '89]   N. Zorboska, *Compact composition operators on some weighted Hardy spaces*, J. Operator Theory **22** (1989), 233–241.

[Zrb '92]   N. Zorboska, *Composition operators with closed range*, preprint, 1992.

Symbol Index

A^2 17

α_p xii, 6, 59

$\angle \lim_{z \to \omega}$ 56

$\approx$ xi

B 14

∂U xi

$\widehat{\mathbf{C}}$ 2

$\chi(T)$ 3

C_φ vii, 11

const. xii

$d(p,q)$ 60

$\Delta(p,r)$ 61

$\ell^1(U)$ 18

ℓ^2 xi

ℓ_G 154

$\ell_U(\gamma)$ 151

$\hat{f}(n)$ xi

$H(\omega, \lambda)$ 62

H^2 9

$H^2(\mathbf{B})$ 20

$HC(T)$ 108

H^p 19

$h_U(z)$ 152

$< \cdot \, \cdot >$ xi

κ xi

$\stackrel{\kappa}{\to}$ xi, 78

k_p 43

L^2 xi

L_α 27

$LFT(\widehat{\mathbf{C}})$ 2

$LFT(U)$ 5

$\mathcal{L}(\mathcal{H})$ 41

$M_2(f,r)$ 12

M_b 11

$\| \cdot \|$ xi, xii

N_φ 179

φ xi

$\varphi'(\omega)$ 56

φ_α 27

φ_n xii, 77

ρ_G 154

$\rho_U(p,q)$ 151

Π xi 6

$\overline{\Pi}$ xi

σ 92, 129

$\mathcal{S}_\varepsilon(\Gamma)$ 158

U xi

$\overline{U}$ xi

Author Index

Abate, M. 87, 203
Ahlfors, L.V. 1, 203
Akeroyd, J. 143, 203
Al'par, L. 20, 203
Aleman, A. 200, 203

Baker, I.N. 144
Bear, H.S. 87
Beardon, A.F. 86, 175, 204
Beauzamy, B. 128, 204
Berkson, E. 200, 204
Birkhoff, G.D.108, 204
Bourdon, P.S. 124, 127, 128,
        141, 143, 145, 204
Brown, J.W. 29, 205
Burckel, R.B. 87, 204
Burdzy, K. 72, 75, 76, 205

Campbell-Wright, R.K. 105, 205
Carathéodory, C. 75, 175,
        197, 205
Carroll, T.F. 76, 205
Caughran, J. 104, 143, 205
Chan, K.C. 127, 205
Churchill, R.V. 29, 205

Cima, J.A. 199, 200, 205
Collingwood, E.F. 197, 205
Cowen, C.C. 104, 143, 144,
        195, 200, 205, 206

Deddens, J. 200, 206
Denjoy, A. 86, 206
Devaney, R.L. 127, 206
Dunford, N. 26, 206
Duren, P.L. 19, 25, 127, 206

Enflo, P. 127, 206
Essén, M. 191, 206

Fan, K. 87, 207
Ford, L.R. 1, 206
Friedman, N. 22, 23, 104
Frostman, O. 183–185, 193,
        197, 207

Gaier, D. 143, 207
Gehring, F.W. 175, 207
Gethner, R.M. 124, 126, 207
Godefroy, G. 124, 126–128, 207
Goebel, K. 86, 207
Goldberg, J.L. 75, 207

Grosse-Erdman, K-G. 128, 207
Guyker, J. 199, 207

Halasz, G. 20, 207
Hayman, W.K. 196, 207
Herrero, D.A. 127, 207
Hervé, M. 75, 87, 207
Herzog, G. 128, 208
Highsmith, J.E. 216
Hunziger, H. 35, 200, 208

Jarchow, H. 35, 200, 208
Julia, G. 61, 63–65, 73, 75,
      82, 84, 208

Königs, G. 89, 90, 101, 103,
      135, 147, 208
Kamowitz, H. 104, 208
Kitai, C. 113, 126, 208
Krantz, S.K. 175, 208
Kreite, T.L. 144, 196, 206, 208
Kubota, Y. 87, 209

Landau, E. 75
Lehner, J. 1, 209
Lindelöf, E. 175, 209
Littlewood, J.E. 19, 38, 197, 209
Lohwater, A.J. 197, 205
Luecking, D. 197, 209
Luh, W. 128, 209

MacCluer, B.D. 20, 52, 53, 87,
      104, 196, 200, 206,
      208, 209
MacLane, G.R. 124, 209
Madigan, K. 200, 209
Marcinkiewicz, J. 128, 209
Markushevich, A.I. 131, 143, 210
Maschioni, V. 200, 208
Menchoff, D. 128, 210

Nevanlinna, F. 75, 210
Nevanlinna, R. 75, 154, 196, 210
Newman, M.H. 23, 210
Nordgren, E.A. 35, 104, 197,
      199, 210

Palka, B.P. 175, 207
Pavone, M. 128, 210
Pommerenke, Ch. 144, 210
Porta, H. 200, 204

Read, C. 127, 210
Reich, S. 86, 207
Riesz, F. 104, 210
Roan, R. 200, 210
Rodin, B. 72, 76, 210
Rolewicz, S. 127, 210
Rosenthal, P. 199, 210
Rudin, W. vii, ix, x, 13, 20, 22,
      25, 29, 30, 39, 41, 49, 75,
      82, 92, 99, 104, 109, 112,
      120, 121, 130–132, 153,
      154, 162, 163, 182, 210
Ryff, J.V. 35, 211

Salas, H. 125, 211
Sarason, D. 75, 200, 201, 211
Schauder, J. 104, 211
Schröder, E. 89, 211
Schwartz, H.J. 35, 104, 211
Schwartz, J.T. 26, 206
Seidel, W. 108, 211
Serrin, J. 75, 211
Shapiro, J.H. 20, 35, 52, 53, 98,
      124, 126, 141, 143, 175,
      195, 200, 204, 205, 207
Shea, D.F. 191, 196, 206
Shields, A.L. 143, 212
Siskakis, A.G. 200, 212
Smith, W. 98, 144, 174, 175, 211
Stanton, C.S. 191, 196, 206, 212
Stegenga, D.A. 98, 144, 174,
      175, 211
Sundberg, C. 196, 200, 212
Sz. -Nagy, B. 104

Taylor, P.D. 35, 53, 196, 211
Thomson, J. 200, 205
Tsuji, M. 72, 212

Ul'yanov, P.L. 128, 212
Ullrich, D.C. 87, 212

Valiron, G. 75, 133, 144, 212
Voronin, S.M. 128, 212

Walsh, J. 108, 211
Wang, Z. 127, 207
Warchawski, S.E. 72, 76, 210
Wintrobe, F.S. 199, 210

Wogen, W. 20, 199, 200, 205, 213
Wolff, J. 75, 86, 212, 213

Young, R.M. 120, 212

Zhu, K. 197, 209
Zorboska, N. 127, 200, 213

Subject Index

α-curve, 45
angular derivative, 56
    and compactness, 56, 57
    and contact, 72
    and tangential
        convergence, 74
    chain rule for, 74
    inadequacy of, 180
    of limit function, 74
    product rule for, 74
angular limit, 56
Arzela-Ascoli Theorem, 22

Bergman space, 17
    compactness problem for, 52
Birkhoff's Theorem, 108
Blaschke
    condition, 181
    product, 181
Bloch
    -'s Theorem, 173
    space, 173
    functions, 173
Burdzy's Theorem, 75

Carathéodory
    Extension Theorem, 130
Change-of-variable Formula, 179
chaotic, 126
Compactness Theorem
    First, 23
    for H^∞, $\ell^1(U)$, 34
    Smooth, 49
    Finitely-Valent, 51
    General, 180
    Hilbert-Schmidt, 26
    Polygonal, 27, 28
    Smooth, 52, 74
    Univalent, 39
Comparison Principle
    for compactness, 31, 59
    for hypercyclicity, 111
composition operator, vii, 11
    compact, 21
    components in space of, 200
    isolation, 200
    semigroups of, 200
    automorphism-induced, 15
    on L^1 and M, 200

on "small" spaces, 200
contact
   α-, 45
   smooth, 45
Contraction
         Mapping Principle, 83
convergence
   uniform on compacts, 34
   weak, 34
curvature, 134, 142
cyclic vector, 107

Denjoy-Wolff Theorem, 78
Denjoy-Wolff point, 78, 85
Dirichlet Space, 18
Disc Convergence Lemma, 62
Distance Lemma, 157
distortion function
   estimates of, 156, 172
   for disc, 152
   for half-plane, 170
   uniqueness of, 172

ε-net, 22
eigenfunctions
   of composition operators, 94

fixed points
   and compactness, 84
   approximate, 82
   attractive, 4
   boundary, 78, 80, 85, 86
   interior, 79
folk-wisdom, 37
Frostman's Theorem, 183

Grand Iteration Theorem, 78
growth estimate
   for H^2 functions, 10

Hardy space H^2, 9
harmonic majorant, 19
harmonic measure, 164
horodisc, 62, 81
hyperbolic
   area, 51, 172

circle, 171
distance, 151
geodesic, 153
length, 151
hypercyclic vector, 107
Hypercyclicity Criterion, 109

intertwining map, 129
Invariant Schwarz Lemma, 60
invariant subspaces, 107, 127, 199
iterates, 77

Jensen's Formula, 118
Jordan domain, 129
Julia's Theorem, 63
Julia-Carathéodory Theorem, 57

Königs
   domain, 147
   function, 92
   Local Theorem, 101
   Theorem, 93, 104, 132
Koebe
   One-Quarter Theorem, 154
   Distortion Theorem, 156
   transform, 156

Lens maps, 27
lens-shaped region, 27, 80
Lindelöf's Theorem, 163
Linear Fractional
   Hypercyclicity Theorem, 114
   Cyclicity Theorem, 121
   Transformation
      conjugates, 2
      derivative, 3
      automorphism, 5
      special automorphism, xii,
         5, 59
      elliptic, 4, 85
      definition of, 1
      fixed points, 2
      hyperbolic, 4
      loxodromic, 4
      matrix representation, 2
      multiplier of, 4

parabolic, 4
standard form 2
trace of, 3
Littlewood's
Inequality, 187
equality in, 197
Subordination
Principle, 13, 19
Theorem, 11, 16
Littlewood-Paley
Identity, 51, 178

MacLane's Theorem, 124
Mergelyan's Theorem, 131
model
Königs's , 147
Jordan, 129
linear fractional, 129
-Theorem, 132
Standard Example, 130
transference from, 129

Nevanlinna Counting Function
definition, 179
sub-averaging property, 190
subharmonic nature, 191
transformation formula, 192
No-Sectors Theorem, 150
proof, 166
non-compactness
via comparison, 31
first example, 30
Improved Theorem, 48
non-tangential limit, 56

operator
-norm, 13
adjoint, 41
algebras, 199
backward shift, 14, 109
bounded, 13
Cesàro, 200
closed range, 199
compact, 21
composition, 11

contraction, 13
cyclic 107
diagonal, 33
finite rank, 22
Hilbert-Schmidt, 26
diagonal, 33
hypercyclic, 107
invertible, 95
multiplication, 11, 13
by z, 110
Schatten class, 196
projection, 22
orbit, 80, 107

pseudo-hyperbolic
disc, 61, 72
distance, 59
triangle inequality, 73

Radial Limit Theorem, 25
reducing subspaces, 199
reproducing kernel, 43
for Bergman space, 52
normalized, 43, 192
ribbon domain, 58
Riesz's Theorem, 95, 104
proof of, 99
Rolewicz's Theorem, 110

Schatten classes, 196
Schröder's equation, 89, 147
Schwarzian deivative, 141
sector, 55
Seidel-Walsh Theorem, 110, 123
similarity of operators, 102, 104
spectrum, 96, 104
Stanton's formula, 196
strictly starlike, 149
subordinate, 32

topological transitivity, 107
Transference Theorem, 130
Tsuji-Warschawski
Theorem, 72, 74
twisted sector, 158
Twisted Sector Theorem, 159

universality, 128

Valiron's Theorem, 133

Walsh's Theorem, 131
weak compactness, 34
Weak Convergence
          Theorem, 29, 34

Wolff point,
     see: Denjoy-Wolff point
Wolff's Theorem, 81

zero-sequence, 118

Universitext *(continued)*

Meyer-Nieberg: Banach Lattices
Mines/Richman/Rultenburg: A Course in Constructive Algebra
Moise: Introductory Problem Course in Analysis and Topology
Montesinos: Classical Tessellations and Three Manifolds
Nikulin/Shafarevich: Geometries and Groups
Øksendal: Stochastic Differential Equations
Porter/Woods: Extensions and Absolutes of Hausdorff Spaces
Rees: Notes on Geometry
Reisel: Elementary Theory of Metric Spaces
Rey: Introduction to Robust and Quasi-Robust Statistical Methods
Rickart: Natural Function Algebras
Rotman: Galois Theory
Rybakowski: The Homotopy Index and Partial Differential Equations
Samelson: Notes on Lie Algebras
Schiff: Normal Families of Analytic and Meromorphic Functions
Shapiro: Composition Operators and Classical Function Theory
Smith: Power Series From a Computational Point of View
Smoryński: Logical Number Theory I: An Introduction
Smoryński: Self-Reference and Modal Logic
Stanišić: The Mathematical Theory of Turbulence
Stillwell: Geometry of Surfaces
Stroock: An Introduction to the Theory of Large Deviations
Sunder: An Invitation to von Neumann Algebras
Tondeur: Foliations on Riemannian Manifolds
Verhulst: Nonlinear Differential Equations and Dynamical Systems
Zaanen: Continuity, Integration and Fourier Theory